AF595340

Advanced Propulsion System and Thermal Management Technology

Advanced Propulsion System and Thermal Management Technology

Editors

Jian Liu
Chaoyang Liu
Yiheng Tong
Bengt Sunden

MDPI • Basel • Beijing • Wuhan • Barcelona • Belgrade • Manchester • Tokyo • Cluj • Tianjin

Editors

Jian Liu
Central South University
Changsha
China

Chaoyang Liu
National University of
Defense Technology
Changsha
China

Yiheng Tong
Space Engineering University
Beijing
China

Bengt Sunden
Lund University
Lund
Sweden

Editorial Office
MDPI
St. Alban-Anlage 66
4052 Basel, Switzerland

This is a reprint of articles from the Special Issue published online in the open access journal *Energies* (ISSN 1996-1073) (available at: https://www.mdpi.com/journal/energies/special_issues/APSTMT).

For citation purposes, cite each article independently as indicated on the article page online and as indicated below:

LastName, A.A.; LastName, B.B.; LastName, C.C. Article Title. *Journal Name* **Year**, *Volume Number*, Page Range.

ISBN 978-3-0365-8004-3 (Hbk)
ISBN 978-3-0365-8005-0 (PDF)

Contents

Editorial

Heat Transfer Enhancement Methods Applied in Energy Conversion, Storage and Propulsion Systems

Wenxiong Xi [1,2], Mengyao Xu [1], Kai Ma [3] and Jian Liu [1,*]

1 Research Institute of Aerospace Technology, Central South University, Changsha 410012, China
2 Hi-Tech Research Institute, Hunan Institute of Traffic Engineering, Hengyang 421001, China
3 Xi'an Modern Chemistry Research Institute, Xi'an 710065, China
* Correspondence: jian.liu@csu.edu.cn

Citation: Xi, W.; Xu, M.; Ma, K.; Liu, J. Heat Transfer Enhancement Methods Applied in Energy Conversion, Storage and Propulsion Systems. *Energies* **2022**, *15*, 7218. https://doi.org/10.3390/en15197218

Received: 13 September 2022
Accepted: 20 September 2022
Published: 1 October 2022

With the development of energy storage and conversion or advanced propulsion systems, heat transfer enhancement methods have become widely applied. For the requirement of clean power, lithium-ion batteries are considered to be the most important choice due to their safety, high-energy density and relatively long working life. Duan et al. [1] conducted a numerical study for the 3D temperature distribution of a lithium-ion battery with a liquid cooling system to ensure suitable working conditions for batteries. In the study, two main effects were considered: the channel size and inlet boundary conditions. They found that the cooling channel width had great effects on the maximum temperature.

Recently, magnetic fields have been widely used in refrigeration and warming applications requiring high heat transmission. For this purpose, magnetic nanofluids have been proposed to enhance heat transfer. Over a linear extending sheet, Hussain et al. [2] investigated the heat transfer characteristics of the magnetohydrodynamic nanofluid flow. In this study, water was the base fluid, with Zn and TiO_2 working as two different types of nanoparticles. This study aimed to investigate the effect of different influential parameters on temperature and flow field distributions for rotating nanofluid in a magnetic field. They found that the rotation could increase the heat flux and decrease friction drag. Another piece of work by this research group concerned rotating nanofluids in an oil-based engine using a two-phase nanoliquid model on an elastic surface [3].

As a renewable energy, concentrated solar power technologies are expected to play a key role in reducing pollution levels. Agyekum et al. [4] investigated two kinds of heat transfer fluids under two power cycles, i.e., supercritical CO_2 and Rankine. Based on the results, the sCO_2 system was proven to be more economically feasible compared to the Rankine system.

Hydrogen is also a viable choice for reducing greenhouse gas emissions, which is a topic that has generated a lot of attention from researchers. There are several methods for the production of hydrogen, including coal gasification, methane cracking, water electrolysis and the partial oxidation of hydrocarbons. Recently, methane pyrolysis in molten metals/salts has been widely investigated, having the advantage of preventing reactor coking and the existence of rapid catalyst deactivation in conventional pyrolysis [5]. Another advantage is the high heat transfer enhancement in the pyrolysis process caused by the large heat capacity of molten media.

Algehyne et al. [6] investigated heat transfer improvements in pseudoplastic materials using ternary hybrid nanoparticles over a stretched porous sheet. The model of energy conservation was built based on a normal heat conduction model focusing on heat generation/absorption. The proposed method was validated by the published findings, and an excellent agreement was found.

Solar energy is considered to be the most appropriate energy source, and has been exploited in different forms, such as in solar air heaters, which realize the conversion of solar energy into thermal air energy. To enhance the thermal performance, Alam at al. [7]

added protrusion ribs to the absorber of solar air heaters. These ribs could enhance the heat transfer of the flowing air without causing a significant pressure drop. The roughness on the absorber could be produced by indenting the device without adding additional mass.

For energy storage applications, the latent heat of phase-change materials is a promising method, although their common disadvantage is their low thermal conductivity, which makes it difficult to use latent heat thermal energy storage. Behi et al. [8] used a tubular heat pipe to improve the heat transfer and increase the charging/discharging rate. The temperature distribution, liquid fraction and charging/discharging rates were discussed. The proposed heat pipe system significantly improved the heat transfer between the fluid and body frame in the operation environment.

Heat transfer enhancement methods are also used in propulsion systems [9] to realize the thermal protection of the engine, especially at high flight speeds. The enhancement of heat transfer in a perforated ribbed channel for blade cooling was investigated by Liu et al. [10]. They utilized steady-state liquid crystal thermography to measure the heat transfer coefficient over ribbed surfaces. It was found that the local heat transfer was enhanced 12–24% in the low heat transfer regions caused by recirculation, and the averaged heat transfer was enhanced by approximately 4–8%. With a small pressure drop penalty, the overall heat transfer performance was improved.

Liu et al. [11] also designed a type of perforated rib with inclined holes for internal cooling passages. This work aimed at furtherly improving the heat transfer of perforated ribbed channels with additional secondary flows caused by inclined hole arrangements. It was found that the penetrated flows mixed strongly with the mainstream flows at the perforated regions for the small inclined angle case. With an increased inclined angle, the penetrated flows approached the inclined direction and mixed with the main stream at the inclined side.

Overall, the enhancement of heat transfer is still widely used in the application of clean energy and energy storage and the conversion, as well as the thermal management of propulsion systems. One kind of heat transfer enhancement is the single-phase flow with increased heat conductivity, such as the use of nanofluids, which can work under certain special working conditions, such as in rotating engines and magnetic fields. Additionally, a convective heat transfer surface can be optimized to search for an improved heat transfer, including shape, flow path and turbulators. Another important method is the use of a two-phase flow, which absorbs plenty of energy during the phase-change process, such as heat pipes in the thermal management of lithium-ion batteries. Research works about the enhancement of heat transfer have been developed along the direction of novel materials, phase changes and multiscale structure designs.

Author Contributions: Conceptualization, W.X.; writing—original draft preparation, J.L.; writing—review and editing, M.X. and K.M. All authors have read and agreed to the published version of the manuscript.

Funding: This work was supported by the National Natural Science Foundation of China (grant no. 12272133).

Conflicts of Interest: The authors declare no conflict of interest.

References

1. Duan, J.B.; Zhao, J.P.; Li, X.K.; Panchal, S.; Yuan, J.L.; Fraser, R.; Fowler, M. Modeling and Analysis of Heat Dissipation for Liquid Cooling Lithium-Ion Batteries. *Energies* **2021**, *14*, 4187. [CrossRef]
2. Hussain, A.; Arshad, M.; Rehman, A.; Hassan, A.; Elagan, S.K.; Ahmad, H.; Ishan, A. Three-Dimensional Water-Based Magneto-Hydrodynamic Rotating Nanofluid Flow over a Linear Extending Sheet and Heat Transport Analysis: A Numerical Approach. *Energies* **2021**, *14*, 5133. [CrossRef]
3. Hussain, A.; Arshad, M.; Rehman, A.; Hassan, A.; Elagan, S.K.; Alshehri, N.A. Heat Transmission of Engine-Oil-Based Rotating Nanofluids Flow with Influence of Partial Slip Condition: A Computational Model. *Energies* **2021**, *14*, 3859. [CrossRef]
4. Agyekum, E.B.; Adebayo, T.S.; Bekun, F.V.; Kumar, N.M.; Panjwani, M.K. Effect of Two Different Heat Transfer Fluids on the Performance of Solar Tower CSP by Comparing Recompression Supercritical CO_2 and Rankine Power Cycles, China. *Energies* **2021**, *14*, 3426. [CrossRef]

5. Msheik, M.; Rodat, S.; Abanades, S. Methane Cracking for Hydrogen Production: A Review of Catalytic and Molten Media Pyrolysis. *Energies* **2021**, *14*, 3107. [CrossRef]
6. Algehyne, E.A.; El-Zahar, E.R.; Sohail, M.; Nazir, U.; Al-bonsrulah, H.A.Z.; Veeman, D.; Felemban, B.F.; Alharbi, F.M. Thermal Improvement in Pseudo-Plastic Material Using Ternary Hybrid Nanoparticles via Non-Fourier's Law over Porous Heated Surface. *Energies* **2021**, *14*, 8115. [CrossRef]
7. Alam, T.; Meena, C.S.; Balam, N.B.; Kumar, A.; Cozzolino, R. Thermo-Hydraulic Performance Characteristics and Optimization of Protrusion Rib Roughness in Solar Air Heater. *Energies* **2021**, *14*, 3159. [CrossRef]
8. Behi, H.; Behi, M.; Ghanbarpour, A.; Karimi, D.; Azad, A.; Ghanbarpour, M.; Behnia, M. Enhancement of the Thermal Energy Storage Using Heat-Pipe-Assisted Phase Change Material. *Energies* **2021**, *14*, 6176. [CrossRef]
9. Liu, C.; Zhang, J.; Jia, D.; Li, P. Experimental and Numerical Investigation of the Transition Progress of Strut-Induced Wakes in the Supersonic Flows. *Aerosp. Sci. Technol.* **2022**, *120*, 107256. [CrossRef]
10. Liu, J.; Hussain, S.; Wang, W.; Xie, G.N.; Sunden, B. Experimental and Numerical Investigations of Heat Transfer and Fluid Flow in a Rectangular Channel with Perforated Ribs. *Int. Commun. Heat Mass Transf.* **2021**, *121*, 105083. [CrossRef]
11. Liu, J.; Hussain, S.; Wang, W.; Wang, L.; Xie, G.N.; Sunden, B. Heat Transfer Enhancement and Turbulent Flow in a Rectangular Channel Using Perforated Ribs With Inclined Holes. *J. Heat Transf.* **2019**, *141*, 041702. [CrossRef]

Article

Generation and Propagation Characteristics of an Auto-Ignition Flame Kernel Caused by the Oblique Shock in a Supersonic Flow Regime

Wenxiong Xi [1], Mengyao Xu [1], Chaoyang Liu [2], Jian Liu [1,*] and Bengt Sunden [1,3]

[1] School of Aeronautics and Astronautics, Central South University, Changsha 410012, China; 13739076081@163.com (W.X.); 15130715317@163.com (M.X.); bengt.sunden@energy.lth.se (B.S.)
[2] College of Aerospace Science and Engineering, National University of Defense Technology, Changsha 410015, China; liuchaoyang08@nudt.edu.cn
[3] Department of Energy Sciences, Lund University, P.O. Box 118, SE-22100 Lund, Sweden
* Correspondence: jian.liu@csu.edu.cn

Abstract: The auto-ignition caused by oblique shocks was investigated experimentally in a supersonic flow regime, with the incoming flow at a Mach number of 2.5. The transient characteristics of the auto-ignition caused by shock evolvements were recorded with a schlieren photography system, and the initial flame kernel generation and subsequent propagation were recorded using a high-speed camera. The fuel mixing characteristics were captured using NPLS (nanoparticle-based planar laser scattering method). This work aimed to reveal the flame spread mechanism in a supersonic flow regime. The effects of airflow total temperature, fuel injection pressure, and cavity length in the process of auto-ignition and on the auto-ignitable boundary were investigated and analyzed. From this work, it was found that the initial occurrence of auto-ignition is first induced by oblique shocks and then propagated upstream to the recirculation region, to establish a sustained flame. The auto-ignition performance can be improved by increasing the injection pressure and airflow total temperature. In addition, a cavity with a long length has benefits in controlling the flame spread from the induced state to a sustained state. The low-speed recirculating region created in the cavity is beneficial for the flame spread, which has the function of flame-holding and prevents the flame from being blown away.

Keywords: auto-ignition; supersonic flow; initial flame kernel; oblique shock; recirculating region

Citation: Xi, W.; Xu, M.; Liu, C.; Liu, J.; Sunden, B. Generation and Propagation Characteristics of an Auto-Ignition Flame Kernel Caused by the Oblique Shock in a Supersonic Flow Regime. *Energies* **2022**, *15*, 3356. https://doi.org/10.3390/en15093356

Academic Editor: Andrey Starikovskiy

Received: 7 April 2022
Accepted: 3 May 2022
Published: 5 May 2022

1. Introduction

Ignition and flame-holding are important issues in aircraft engine applications such as gas turbine combustors, ramjets, and scramjets. In these high-speed flow fields, the principle of combustion organization is to maintain a velocity equivalency relationship between the incoming reactant mixture and upstream flame propagation. The traditional flame propagation theory mainly concentrates on explaining the mechanism of flame-holding [1,2]. However, for a scramjet, the auto-ignition occurs when the air flow temperature increases to a threshold during the process of cruise flight. Fuel auto-ignition plays an important role in the process of velocity balance between the incoming reactant mixture and upstream flame propagation [3,4], which greatly affects the performance of the engine.

As a fundamental problem in combustion, auto-ignition phenomena have generally been investigated under various flow conditions in the past years [5,6]. Numerical calculations were used by different researchers to describe flame behaviors [7–9]. The chemical kinetics of auto-ignition, ignition delay time, and key mechanism were studied using shock tubes with engine-like conditions [10–12]. Meanwhile, the auto-ignition behavior within high temperature air is generally investigated using classical and simplified experimental facilities, such as a co-flow jet flame [13–15] and counter-flow flame [16,17]. In these studies, the kinetic-controlled auto-ignition phenomena are described in view of laminar and diffusion flames. However, the supersonic flow characteristics in scramjets, such as shock waves, have not been well considered, and the effect of shock waves on auto-ignition has been ignored. Recently, a more scramjet-like experimental condition was tested, and the flame structure in the crossflow of the heated air was observed, to analyze the behavior of auto-ignition and lift-off in a subsonic flow [18,19]. The results showed the process of auto-ignition for sustained flames in a high-temperature condition. However, how the auto-ignition and flame holding functions in the supersonic flow regime is still an unresolved problem.

The complexity of ignition and flame-holding in a supersonic flow-field is caused by the introduction of additional flow phenomena, such as shock waves, a strong shear layer, unstable wall boundary, and flow recirculation [20–22]. Although ignition and flame-holding have been investigated by some researchers, the mechanism based on the quasi-steady assumption of combustion is not available for the ignition transient, because of the interference between the shock and the flame. During the ignition transient, the generation of a shock wave has obvious influences on the combustion heat release and pressure increase downstream. The shock wave induced by combustion can change the fuel distribution from the initial cold condition and stationary combustion condition. Thus, the results obtained with a cold flow and stable combustion cannot be simply applied to common ignition dynamics [23–25]. Meanwhile, a shock wave has been proven to play a key role in the process of changing from self-ignition, transition, and flame-holding. It is found that the transition to flame-holding is both mixing-limited and reaction-limited, because of the shock wave, which increases the Damköhler number Da significantly. When sustained combustion is established, the combustion mode transition from a ramjet to a scramjet still depends on the state of the shock train structure [26]. How to distinguish the flame connecting phenomena caused by the co-existence of high-speed core flows and low velocity recirculating flows remains a problem. The occurrence of the flame depends on the fuel injection, flame holding device, the state of incoming air, and combustor structures. Rasmussen [27] investigated the flame stabilization mechanism using laser-induced fluorescence of OH and CH2O within a free stream with Ma = 2.4. The effect of the arrangement of fuel injections on combustion was studied. They found that the placement of the fuel injector had significant effects on the flame-holding mechanisms in a directly-fueled cavity flame-holder.

However, information on auto-ignition flame spread in the supersonic flow regime, particularly the effect of recirculation and oblique shocks on flame kernel generation and propagation, has rarely been covered in detail. In the present work, the auto-ignition process of hydrogen and ethylene was investigated experimentally in a supersonic flow regime with incoming air at Mach = 2.5. First, the auto-ignitable boundary was sketched, based on numerous ignition tests under various air flow total temperatures and fuel injection pressures. Then, the process of the generation of an initial flame kernel and propagation was observed with a high-speed camera and a schlieren photography system. Finally, the effect of the length of the cavity on flame establishment and propagation is considered and the auto-ignition mechanism is discussed.

2. Experimental Approach

2.1. Experimental Facility

The auto-ignition experiments were conducted in a directly-connected supersonic combustion test facility, which was composed of an air heater, an isolator, a combustor, and an expanding sector. Figure 1 shows a schematic of the layout of the ignition test system. The mass flow rate in the supersonic combustor was 1 kg/s at Ma = 2.5, and the total pressure was 1.6 MPa. The incoming air was heated through a vitiated heater, in which ethanol, oxygen, and air were supplied to burn and accordingly generate heat. The entrance condition was set based on the typical cruise flight of a scramjet. The total temperature of the heated air could be changed by adjusting the heat release power in the vitiated heater. More details about the ignition facility are described in refs. [28,29].

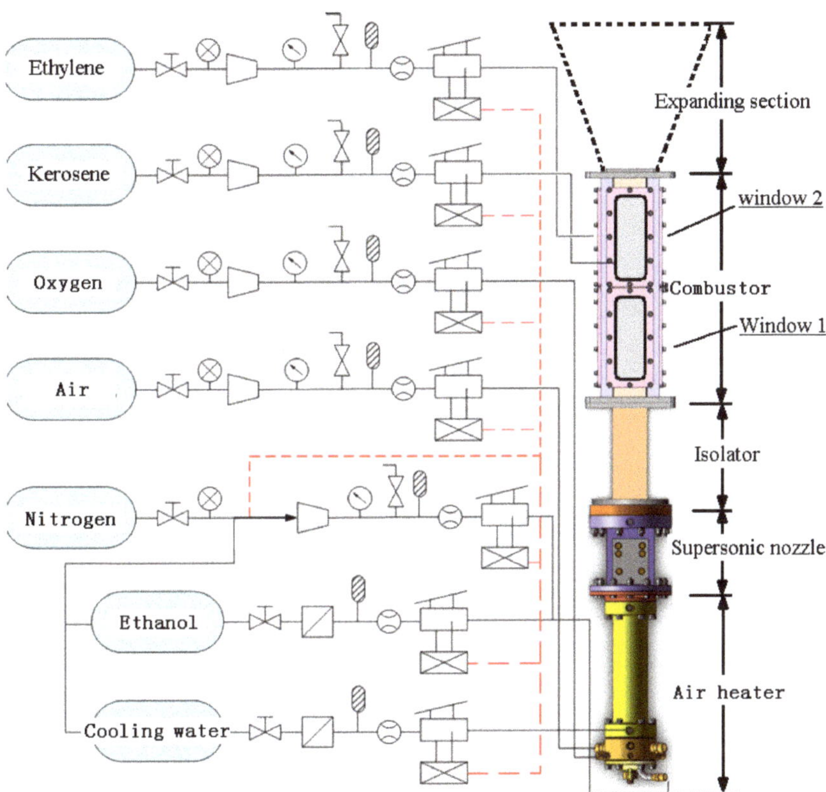

Figure 1. Schematic of the ignition system.

2.2. Combustor Configuration

Figure 2 shows a schematic diagram of the supersonic combustor for the auto-ignition test. The height of the straight combustor section was 40 mm, and the width was 50 mm. It was connected to an expanding section for exhaust gas. The cavity used for flame-holding had a depth of 8 mm. The cavity length could be changed from 24 mm to 56 mm, to investigate the effect of the cavity length. The cavities with different lengths were, respectively, named LD3, LD5, and LD7. Ethylene and hydrogen were used as the test fuel. The fuel was injected at sonic speed in a direction perpendicular to the air stream through a single hole with a diameter of 1 mm, located in the test section wall. The distance from the injection placement to the cavity was 30 mm. The fuel injection pressure could be adjusted from 1 MPa to 3.2 MPa, with a fuel stagnation temperature of 296 K. The total temperature of the incoming air was varied from 1400 K to 1650 K. Upstream of the cavity, at a distance of 60 mm, a pressure sensor was installed to measure the combustor pressure. To improve the auto-ignition performance of the fuel, a ramp with an inclination angle of 18 degrees was installed on the top wall to generate oblique shocks. Related geometric parameters and flow conditions are listed in Table 1.

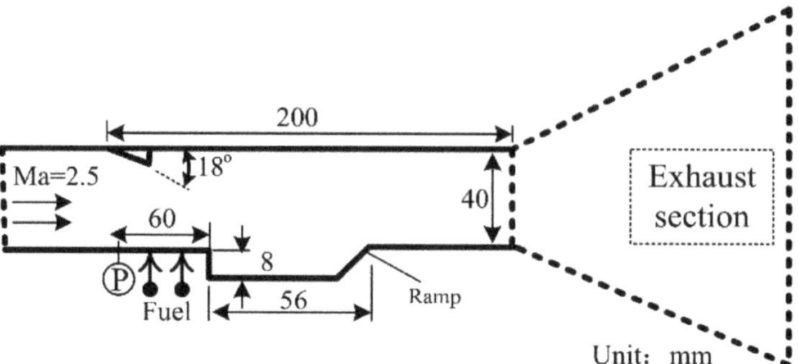

Figure 2. Schematic diagram of the supersonic combustor for the auto-ignition test.

Table 1. Related parameters of the supersonic combustor.

Related Parameters			
Mach number	2.5	Total temperature	1400–1650 K
Length of fuel injection	60 mm	Injection pressure	1.0–3.2 Mpa
Length of cavity	24–56 mm	Angle of oblique shock	18°
Depth of cavity	8 mm	Height of combustor	40 mm
Width of combustor	50 mm	Test length of combustor	200 mm

2.3. *Method for Flow-Field and Flame Observation*

Quartz windows were mounted on the side walls of the combustor, to enable optical measurements. Schlieren photography, as shown in Figure 3c, was performed using a high speed video system MotionBLITZ (Mikrotron GmbH, Unterschleißheim, Germany) and was used to record the shock train involution during the ignition transient. A 532-nm laser system was used as a back light source, to eliminate the interference of the flame emission. Images of the initial auto-ignition flame kernel and subsequent flame propagation were acquired with a Photron FASTCAM-SA5, as shown in Figure 3b. The camera system consisted of a vidicon, lens, and processor (Photron Limited, San Diego, CA, USA). The lens was a Nikon 50~500-mm zoom lens (Photron Limited San Diego, CA, USA). The image sensor was a CMOS with high sensitivity, and the frame rate could reach 120 thousand frames per second.

The mixing characteristics of the vertical fuel injection were displayed utilizing NPLS (nanoparticle-based planar laser scattering) technology. The NPLS measurement system consisted of a computer, synchronizer, CCD camera, and pulse laser, as shown in Figure 3d. In the NPLS measuring system, the computer controlled the interaction of the components and received the experimental images. The input and output parameters of the synchronizer were controlled by software, and the related assisting-computers were controlled by the signal of the synchronizer. Time diagrams of the CCD exposure and laser output of the pulse laser were adjusted for the purpose of measurement. The laser beam was transformed into a sheet using a cylindrical lens. In the experiment, the thickness of the laser sheet was 1 mm. The synchronizer had eight output ports, with a time accuracy of 0.25 ns. The CCD camera was an interline transfer type CCD, with a shortest frame time of 200 ns and array resolution of 2000 × 2000. The gray levels of each pixel were 4096, using a micro-lens. The pulsed laser light source was a double cavity Nd: YAG laser machine with an output laser wavelength of 532 nm, pulse duration time of 6 ns, and highest energy of a single pulse of 350 mJ.

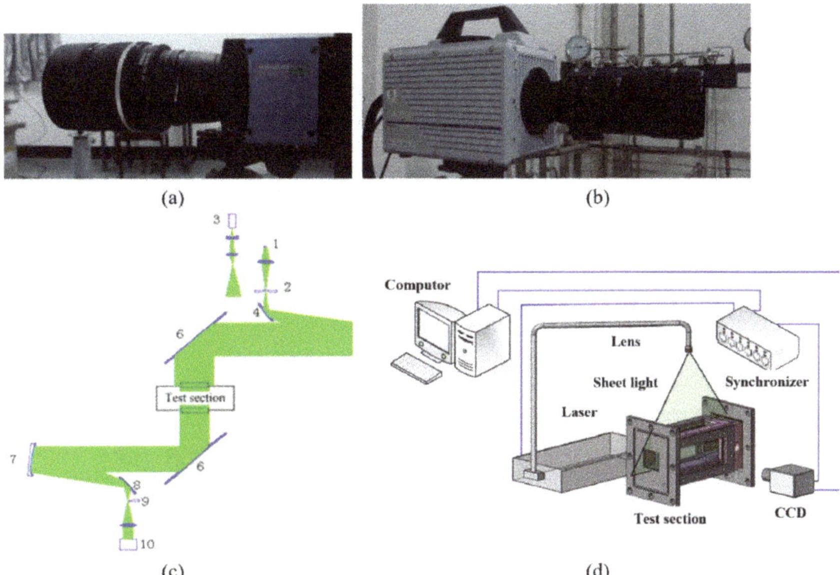

Figure 3. High speed image recording devices used for the ignition tests: (**a**) MotionBLITZ; (**b**) Photron FASTCAM-SA5; (**c**) set-up of the schlieren photography; (**d**) set-up of the test section and the NPLS system.

3. Results and Discussion

3.1. Results of Auto-Ignitable Boundary

In the current work, the occurrence of auto-ignition was captured from the light scattering images through the windows. The ignition process began with a well-established incoming flow condition with the air heater. Then the fuel injection switch was turned on, with a duration of 2 s, which is a sufficiently long time to reach a time-averaged non-premixed flow field. Auto-ignition took place, with the occurrence of an initial flame kernel. Based on the acquired sequential flame spread images, three cases of auto-ignition event could be classified: a failed case, without initial flame kernel generation; a case of auto-ignition kernel appearance, without a sustained flame; and a case of successful global ignition. To investigate the effect of the total temperature and injection, several ignition tests were performed, with a stepwise increase of temperature by 25 K and pressure by 0.5 MPa.

Figure 4 presents the experimental results of the auto-ignitable boundary under different injection pressures and total temperatures. The suitable auto-ignitable area is shown in the right corner of the figure. From the figure, it is evident that a higher total temperature of the incoming air and injection pressure improved the performance of auto-ignition. A high temperature reduced the auto-ignition delay time of the reactants, and a high injection pressure enhanced the fuel mixing and decreased the mixing time, to create an ignitable concentration condition. Overall, auto-ignition appeared in a temperature range of 1500 K–1650 K and injection pressure range of 2.0 MPa–3.5 MPa. From the experimental results, it can be seen that total temperature had a greater influence compared to the injection pressure. Auto-ignition could occur at a high total temperature of incoming air, even with a relatively low injection pressure. From the auto-ignitable boundary with a "C" shape, it is also evident that for a certain total temperature of incoming air, there exists an optimized injection pressure of fuel.

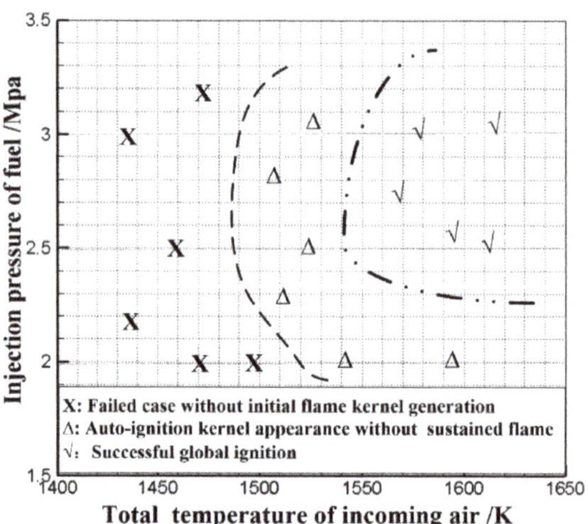

Figure 4. Experimental results of auto-ignitable boundary under different injection pressures of the ethylene and total temperatures of the incoming air.

3.2. *Auto-Ignition Transient and Initial Flame Kernel Propagation*

The laser schlieren system was employed to acquire the auto-ignition transient. Figure 5 presents typical sequential schlieren images of the ethylene auto-ignition and the shock wave evolvement of the auto-ignition. The total temperature of the incoming air was 1480 K and the injection pressure of the ethylene was 1.5 MPa. In Figure 5, the layout and interaction of shock waves are clearly displayed. The mixing of the fuel injection along the combustor can be approximately evaluated according to the faint contrast grade. It is noted that the boundary edge of fuel transportation has been manually marked with dash-dotted lines. Figure 5a shows the initial phase of the shock evolution. The time interval of the snapshots was 0.4 ms, with an exposure time of 50 μs. The subsequent quasi-stable phase of the establishment of the shock train is depicted in Figure 5b, with a time interval of 0.2 ms. The first image of Figure 5a displays the flow-field characteristics just before the emergence of auto-ignition. As shown in the photograph, several typical shocks are clearly observed: an oblique shock induced by the ramp, a bow shock by the fuel injection, the expanding shock of the ramp rear, and the intersection reflected by the shear layer. When the ignition begins, the intersection circled with dashed lines in the figure is pushed ahead, and then the shock train forms. The process of shock train generation was accompanied by a boundary separation caused by downstream heat release. Compared with the cold condition before ignition, the intensity of the fuel mixing in the state of stable combustion was enhanced due to the heat release. One reason for this is that the depth of fuel penetration was increased due to the decreased velocity of the main-flow. Another reason is that the turbulence induced by the shock train also improved the mixing of the reactants.

Figure 6 illustrates corresponding dynamic flame sequential images of the ethylene auto-ignition. The time interval of the snapshots was 83 μs. As shown in Figure 6, the process of flame propagation is divided into two phases: the initial flame kernel generation phase (left), and the quasi-stable flame phase (right), corresponding to Figure 5a,b. The first phase involved the initial flame generation and heat accumulation by the downstream strong combustion. The second phase refers to the upstream flame spread through the low speed circulation cavity and the establishment of a sustained flame.

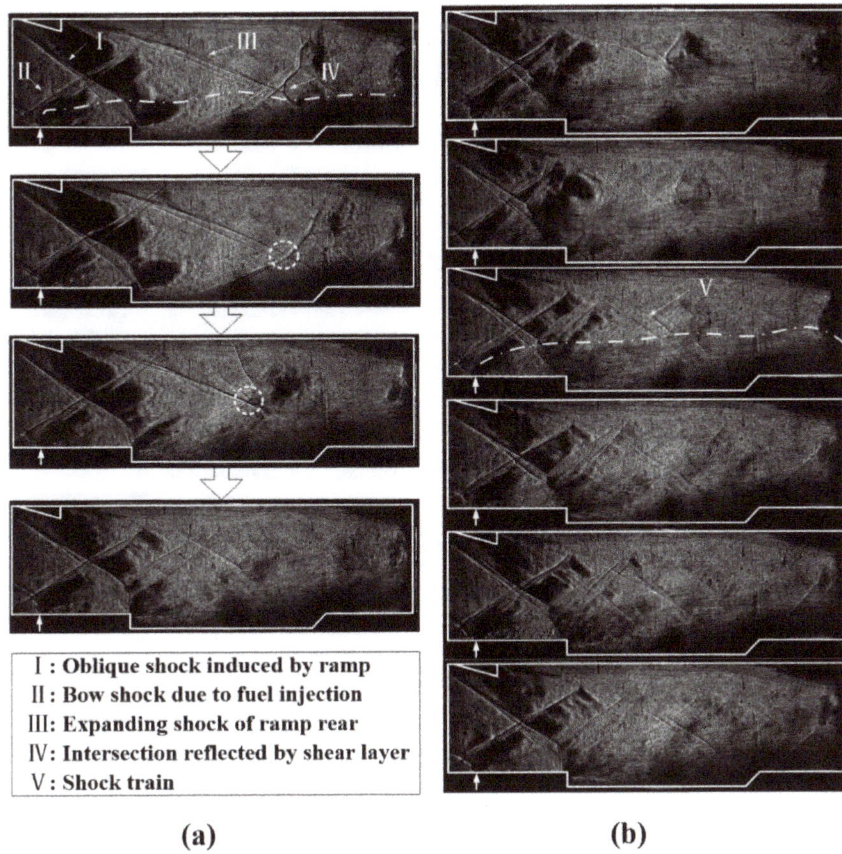

Figure 5. Sequential schlieren images of the ethylene auto-ignition: (**a**) initial phase of the shock evolution, with a time interval of 0.4 ms and an exposure time of 50 μs; (**b**) subsequent quasi-stable phase of the establishment of the shock train, with a time interval of 0.2 ms and an exposure time of 50 μs.

By analysis of the shock layout displayed in Figure 5, it was revealed that the initial flame kernel was caused by the incidence of the oblique shock on the ramp. The reason for this is that the auto-ignition delay time of the reactants was shortened due to the increased temperature and pressure behind the shock. The first visible auto-ignition kernel was at an angle relative to the core of the flow field just downstream of the cavity. The distance from the fuel injection point to the first visible flame kernel was defined as the ignition delayed distance $L_{ignition}$, with a length of 76 mm. Unlike the case of the combustion directly induced by the shock, this made the interaction between the flame and the shock uncoupled, with a relatively long delayed distance. The feedback of the ignition transition and flame propagation was caused by the downstream accumulation of heat release. The heat release made the pressure increase to the boundary separation, and then a shock train was formed. The effect of the formed shock train on the flame propagation is displayed in the seventh image of Figure 6, in which the edges of the flame were created by the shock train.

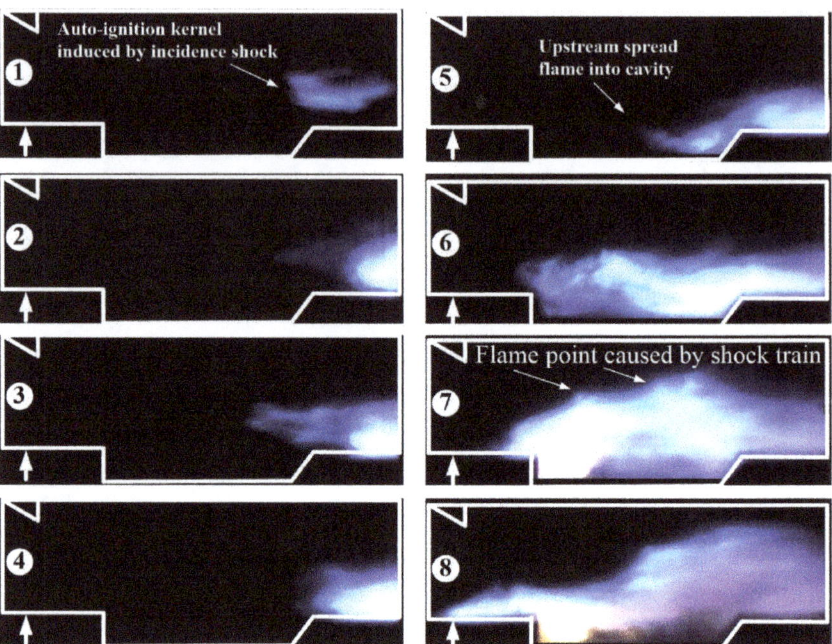

Figure 6. Dynamic flame sequential images of the ethylene auto-ignition, with a time interval of 83 μs.

3.3. Effect of the Cavity on Flame Stabilization and Propagation

To investigate the influence of the cavity on the establishment of a sustained auto-ignition flame, the generation and propagation of the ethylene auto-ignition flame without the cavity was measured as a baseline case. Dynamic and sequential flame images of the ignition are shown in Figure 7, in which the ramp was present to create an oblique shock. Similar to the case of Figure 6, the flame kernel was located at the core of the flow-field, with a certain ignition delay. The subsequent flame propagation followed the common flame spread routine through the separated boundary. The characteristics of the flame dynamics play an important role in this phenomenon. As shown in Figure 7, the flame shape changed abruptly and changed with a low frequency. The unsteadiness is also clearly evident from the pressure response in Figure 8.

Figure 8 shows the pressure response characteristics of the ethylene auto-ignition process. The pressure measured near the wall of the combustor is indicated by Pen, and it varied from 90 kPa to 150 kPa, with time intervals ranging from 115 ms to 296 ms. The total pressure of the main flow (P_c) measured in the air heater and the injection pressure of ethylene (P_{c2h4}) were almost constant during the auto-ignition period. Therefore, the unsteadiness of the airflow and the fuel supply system was not caused by the variation of the pressure response. It can be concluded that the shock generated by the ramp was the main factor behind the occurrence of the auto-ignition flame kernel, and the flow recirculation created by the cavity had benefits for the flame's upstream propagation and stabilization.

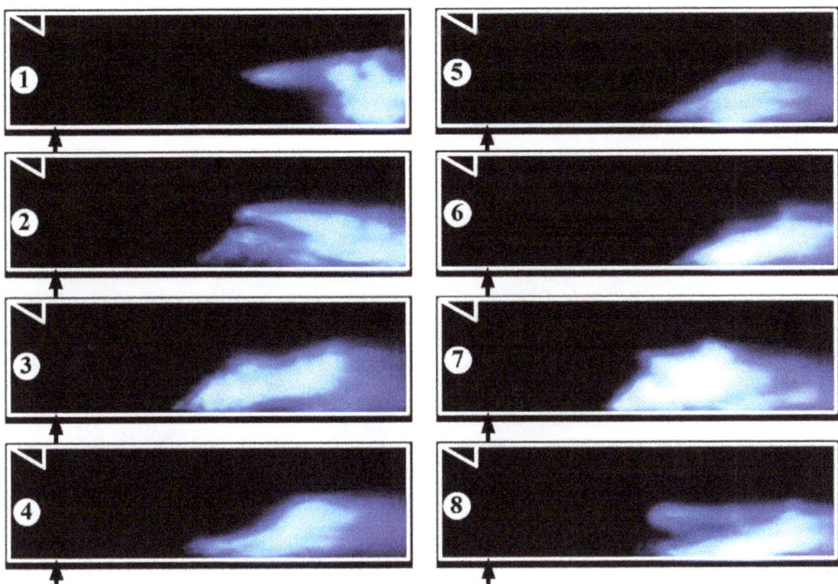

Figure 7. The dynamic flame images of the ethylene auto-ignition without the cavity (the image parameters are similar to Figure 5).

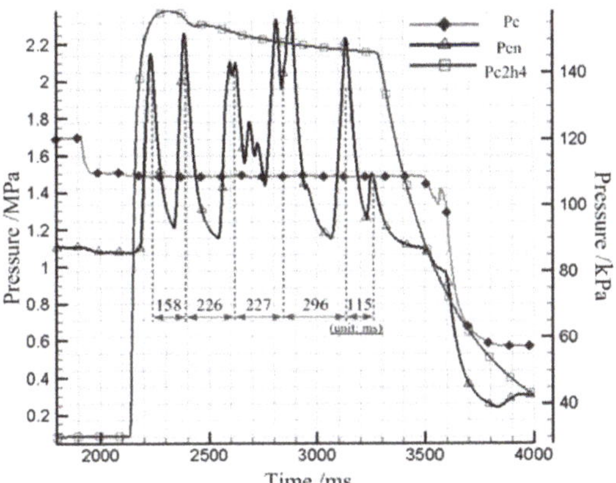

Figure 8. Pressure response characteristics of the ethylene auto-ignition process (P_c indicates pressure of the air heater; P_{en} indicates pressure measured in the supersonic combustor; P_{c2h4} indicates injection pressure of the ethylene).

To obtain more details about the effect of the cavity on flame propagation and stabilization, three types of cavities were tested and compared. Considering the ignition efficiency, hydrogen was selected as the test fuel. Compared with ethylene, the ignition delay time for hydrogen is shorter, with a fast flame spread velocity in the same conditions. Owing to the weak radiation of the hydrogen flame, heated air was seeded with nanoparticles of SiO2 (silicon dioxide) to intensify the flame visibility, which provided benefits in acquiring photographs with a much shorter exposure time and higher acquisition speed, for capturing the instantaneous flame dynamics. Figure 9 presents the initial flame kernel evolvement of the hydrogen auto-ignition. Three types of cavities were compared, with various cavity lengths, i.e., LD7, LD5, and LD3. The injection pressure was set as 2.5 MPa. As depicted in Figure 9, the characteristics of flame generation and propagation were similar to the case described above. The length of the cavity had no obvious influence on the location of the occurrence of the flame kernel. Thus, the delayed residence time of the reactant by the cavity was not the main mechanism behind the first occurrence of auto-ignition. In addition, the extended recirculation zone with a longer cavity is beneficial for the initial heat release accumulation during the flame upstream propagation. The LD7 cavity had the strongest heat release, with an intensive light emission; while the flame of the shorter cavity tended to be blown away.

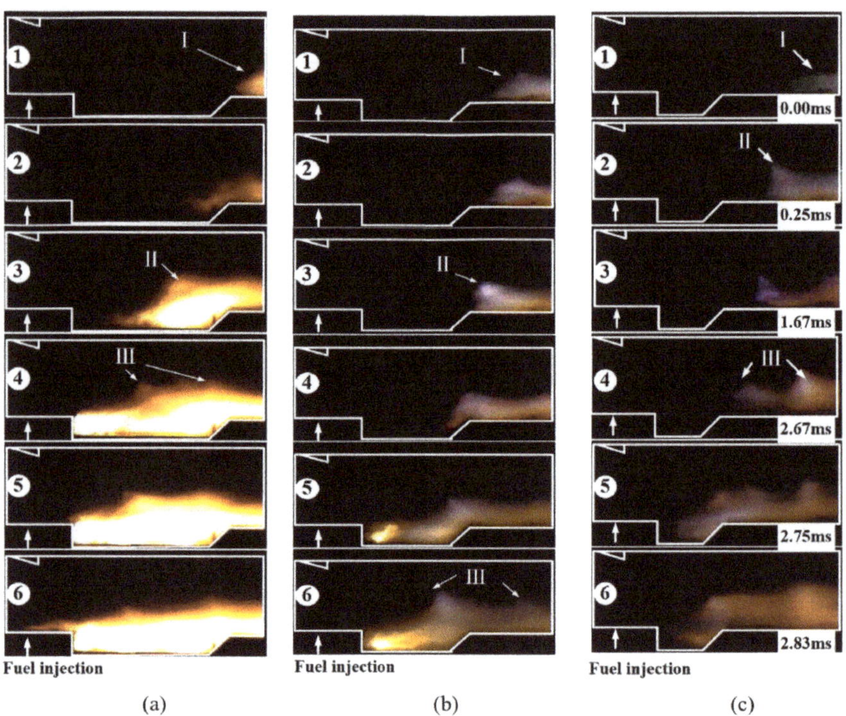

Figure 9. Initial flame kernel evolvement of the hydrogen auto-ignition with various cavity lengths: (**a**) case of LD7 cavity; (**b**) case of LD5 cavity; (**c**) case of LD3 cavity (the exposure times were all set as 83 μs, with a frame speed of 12,000 f/s, except for (**c**), whose time intervals are marked on the images; I indicates the occurrence of the initial flame; II indicates the flame position induced by the incidence shock; III indicates the flame point caused by the shock train).

3.4. Fuel Distribution Based on NPLS Method

Figure 10 shows the mixing characteristics of the fuel injection observed using NPLS when a steady flame was established. The nanoparticle generator was driven by high-pressure gaseous ethylene. When the flow field was measured with NPLS, the nanoparticles were injected into, and mixed with, the inflow of the flow field. When the flow was established in the observation window, the synchronizer controlled the laser pulse and the CCD camera, to ensure synchronization of the emissions scattered by nanoparticles and the exposure of the CCD. The scattering image gave a clear fuel distribution boundary, in which the fuel penetration depth was influenced by the injection pressure. Combined with the flame image obtained above, this shows that a delay in the distance existed for the mixing between the fuel and the incoming air. The front flame edge could propagate upstream along a suitable concentration boundary, until a steady flame was formed.

Figure 10. Mixing characteristic of fuel injection observed using NPLS.

3.5. Analysis of Auto-Ignition and the Flame Spread Mechanism

Based on the aforementioned analysis, a fundamental physical description of the occurrence of auto-ignition and flame propagation can be summarized. The auto-ignition process with a cross-flow effect is shown in Figure 11. After the initial flame kernel was formed, it was divided into two parts and transported downstream. One part was further mixed and burned with the mainstream flame kernel that had not completed self-ignition and accumulated heat, which is represented by propagation path 1. The other part underwent mixing combustion in the low-speed region in the boundary layer, represented by propagation path 2. When the combustion heat release caused boundary layer separation, a boundary layer countercurrent flame appeared. At this time, the flame propagation path 3 spread upstream along the equivalent of the chemically appropriate equivalent ratio, presenting the characteristics of a "three-bifurcation flame", which is stable at a special position. For a successful auto-ignition event, there are three sub-processes required: the mixing process, induction process, and thermal runaway process. The mixing process, similarly to the preparation process, includes fuel injection and transport downstream without chemical reactions. The distance for forming a suitable concentration condition is defined as the mixing length indicated by L_{mix}. Subsequently, the distance of the induction process is named the induction length (L_{induce}), in which the primary reaction process starts, together with the accumulation of the auto-ignition precursor formaldehyde (CHO). The existence of an induction length makes the flame front invisible. The subsequent thermal runaway process is characterized by a massive heat release, accompanied with a visible light scattering of OH radiation. The following two sub-processes are recognized as distributed auto-ignition reactions and disturbed reaction layers from the perspective of stationary flame structures [18]. At the ignition transient, a complete flame spread involves the downstream convective and diffusive flame, and the upstream boundary flame.

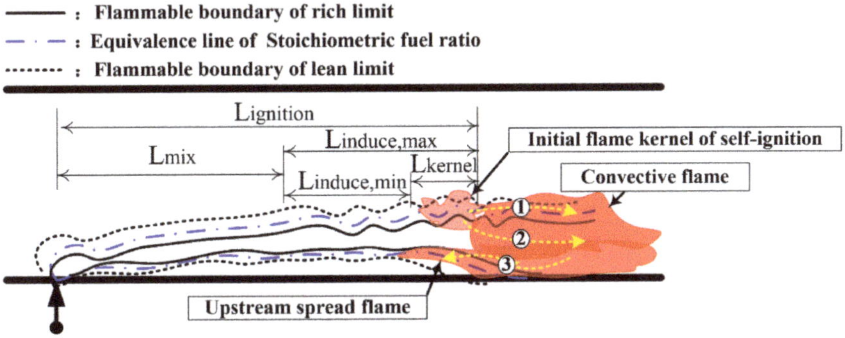

Figure 11. Schematic diagram of auto-ignition with cross-flow injection.

Figure 12 explains the occurrence of flame auto-ignition and the spread mechanism in a supersonic flow with the placement of the ramp and the cavity. From the figure, it is seen that 4, 5, 6, and 7 are developments of path 3; i.e., upstream spread flame in the ramp and the cavity. While, 4 and 5 are the upstream spread flame in the cavity, 6 is the trapped flame with flow circulation and 7 is the downstream spread flame in the cavity. In Figure 12, the incidence of the shock can reduce the mixing length and the induction length. Thus, the induction reaction is brought forward. When the initial flame kernel is generated, the downstream flame development is similar to the case mentioned above in Figure 11. It is necessary to span an additional distance for the combustion process, shown as L_{spread}. The role of the cavity is to reduce the L_{spread} and ensure upstream propagation, until the trapped circulation flame is established. A cavity with a long length has benefits for the flame transition from the induced state to a sustained state.

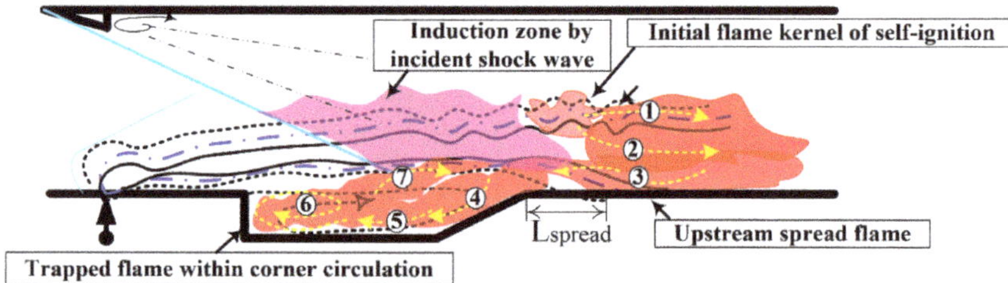

Figure 12. Schematic diagram of an auto-ignition flame and spread mechanism, with the effects of the ramp and the cavity.

4. Conclusions

In the present work, auto-ignition of ethylene and hydrogen caused by an oblique shock was investigated experimentally in a supersonic flow regime, with incoming air at Mach = 2.5. The auto-ignition transient was observed using a schlieren photography system, and the process of the occurrence of the flame kernel and its subsequent propagation was acquired using a high-speed camera. The effects of the incoming airflow total temperature, fuel injection pressure, and length of the cavity on the auto-ignition and auto-ignitable boundary were considered and compared. Several conclusions were drawn, as follows:

(1) The auto-ignition performance was extended by increasing the injection pressure and total temperature of the airflow, with enhanced mixing efficiency and decreased ignition delay time.

(2) The flame evolvement was described as follows: the occurrence of auto-ignition is induced by oblique shocks and then propagates upstream to the recirculating region. Finally, a sustained flame is established inside the recirculating region.

(3) A cavity with a long length has benefits for the flame transition from the induced state to a sustained state. Meanwhile, the low speed recirculation created in the cavity has benefits for the flame spread, with the function of preventing the flame from being blown away by flow oscillations.

Author Contributions: Conceptualization, W.X. and J.L.; methodology, W.X.; software, M.X.; validation, W.X., J.L. and C.L.; investigation, W.X.; writing—original draft preparation, J.L.; writing—review and editing, B.S.; supervision, C.L.; funding acquisition, W.X. All authors have read and agreed to the published version of the manuscript.

Funding: This work was supported by the National Natural Science Foundation of China (Grant No. 51606220).

Conflicts of Interest: The authors declare no conflict of interest.

Nomenclature and Abbreviations

Latin characters	
Da	Damköhler number
$L_{ignition}$	ignition delayed distance
L_{induce}	induction length
L_{mix}	mixing length
L_{spread}	flame spread length
Ma	Mach number
P_{en}	pressure in the combustion chamber
P_c	total pressure of the air flow
P_{c2h4}	injection pressure of the ethylene
Abbreviations	
CCD	charge coupled device
CMOS	complementary metal oxide semiconductor
NPLS	nanoparticle-based planar laser scattering

References

1. Chung, S.H. Stabilization, propagation and instability of tribrachial triple flames. *Proc. Combust. Inst.* **2007**, *31*, 877–892. [CrossRef]
2. Buckmaster, J. Edge-flames. *Prog. Energy Combust. Sci.* **2002**, *28*, 435–475. [CrossRef]
3. Gordon, R.L.; Masri, A.R.; Pope, S.B.; Goldin, G.M. Transport budgets in turbulent lifted flames of methane autoigniting in a vitiated co-flow. *Combust. Flame* **2007**, *151*, 495–511. [CrossRef]
4. Liu, C.; Zhang, J.; Jia, D.; Li, P. Experimental and numerical investigation of the transition progress of strut-induced wakes in the supersonic flows. *Aerosp. Sci. Technol.* **2021**, *120*, 107256. [CrossRef]
5. Mastorakos, E. Ignition of turbulent non-premixed flames. *Prog. Energy Combust. Sci.* **2009**, *35*, 57–97. [CrossRef]
6. An, B.; Sun, M.; Wang, Z.; Chen, J. Flame stabilization enhancement in a strut-based supersonic combustor by shock wave generators. *Aerosp. Sci. Technol.* **2020**, *104*, 105942. [CrossRef]
7. Jones, W.P.; Navarro-Martinez, S. Numerical study of n-Heptane auto-ignition using LES-PDF methods. *Flow Turbul. Combust.* **2009**, *83*, 407–423. [CrossRef]
8. Jones, W.P.; Navarro-Martinez, S.; Rohl, O. Large eddy simulation of hydrogen auto-ignition with a probability density function method. *Proc. Combust. Inst.* **2007**, *31*, 1765–1771. [CrossRef]
9. Richardson, E.S.; Yoo, C.S.; Chen, J.H. Analysis of second-order conditional moment closure applied to an auto-ignitive lifted hydrogen jet flame. *Proc. Combust. Inst.* **2009**, *32*, 1695–1703. [CrossRef]
10. Gokulakrishnan1, P.; Gaines, G.; Klassen, M.S. Autoignition of Aviation Fuels: Experimental and Modeling Study. In Proceedings of the 43rd AIAA/ASME/SAE/ASEE Joint Propulsion Conference & Exhibit, Cincinnati, OH, USA, 8–11 July 2007.
11. Dean, A.J.; Penyazkov, O.G.; Sevruk, K.L.; Varatharajan, B. Autoignition of surrogate fuels at elevated temperatures and pressures. *Proc. Combust. Inst.* **2007**, *31*, 2481–2488. [CrossRef]
12. Athithan, A.A.; Jeyakumar, S.; Sczygiol, N.; Urbanski, M.; Hariharasudan, A. The Combustion Characteristics of Double Ramps in a Strut-Based Scramjet Combustor. *Energies* **2021**, *14*, 831. [CrossRef]
13. Medwell, P.R.; Kalt, P.A.M.; Dally, B.B. Imaging of diluted turbulent ethylene flames stabilized on a Jet in Hot Coflow (JHC) burner. *Combust. Flame* **2008**, *152*, 100–113. [CrossRef]
14. Oldenhof, E.; Tummers, M.J.; van Veen, E.H.; Roekaerts, D.J. Ignition kernel formation and lift-off behaviour of jet-in-hot-coflow flames. *Combust. Flame* **2010**, *157*, 1167–1178. [CrossRef]
15. Kaiser, S.A.; Frank, J.H. Spatial scales of extinction and dissipation in the near field of non-premixed turbulent jet flames. *Proc. Combust. Inst.* **2009**, *32*, 1639–1646. [CrossRef]
16. Pellett, G.L.; Vaden, S.N.; Wilson, L.G. Opposed Jet Burner Extinction Limits: Simple Mixed Hydrocarbon Scramjet Fuels vs. Air. In Proceedings of the 43rd AIAA/ASME/SAE/ASEE Joint Propulsion Conference & Exhibit, Cincinnati, OH, USA, 8–11 July 2007.
17. Ruan, C.; Chen, F.E.; Yu, T.; Cai, W.W.; Li, X.L.; Lu, X.C. Experimental study on flame/flow dynamics in a multi-nozzle gas turbine model combustor under thermo-acoustically unstable condition with different swirler configurations. *Aerosp. Sci. Technol.* **2020**, *98*, 11. [CrossRef]
18. Micka, D.J.; Driscoll, J.F. Stratified jet flames in a heated (1390 K) air cross-flow with autoignition. *Combust. Flame* **2012**, *159*, 1205–1214. [CrossRef]
19. Kurdyumov, V.; Fernandez-Tarrazo, E.; Linan, A. The anchoring of gaseous jet diffusion flames in stagnant air. *Aerosp. Sci. Technol.* **2002**, *6*, 507–516. [CrossRef]
20. Masuya, G.; Choi, B.; Ichikawa, N. Mixing and combustion of fuel jet in pseudo-shock waves. In Proceedings of the 40th AIAA Aerospace Sciences Meeting & Exhibit, Reno, NV, USA, 14–17 January 2002.
21. Mai, T.; Sakimitsu, Y.; Nakamura, H.; Ogami, Y.; Kudo, T.; Kobayashi, H. Effect of the incident shock wave interacting with transversal jet flow on the mixing and combustion. *Proc. Combust. Inst.* **2011**, *33*, 2335–2342. [CrossRef]

22. Li, F.; Sun, M.; Cai, Z.; Chen, Y.; Sun, Y.; Li, F.; Zhu, J. Effects of additional cavity floor injection on the ignition and combustion processes in a Mach 2 supersonic flow. *Energies* **2020**, *13*, 4801. [CrossRef]
23. Tetsuo, H.; Kan, K.; Sadatake, T.; Izumikawa, M.; Watanabe, S. Gas-sampling survey from exhaust flows in scramjet engines at Mach-6 flight condition. In Proceedings of the 40th AIAA/ASME/SAE/ASEE Joint Propulsion Conference and Exhibit, Fort Lauderdale, FL, USA, 11–14 July 2004.
24. Dai, J.; Cai, G.B.; Zhang, Y.; Yu, N.J. Experimental and numerical investigation of combustion characteristics on GO2/GH2 shear coaxial injector. *Aerosp. Sci. Technol.* **2018**, *77*, 725–732. [CrossRef]
25. Jeong, E.; O'Byrne, S.; Jeung, I.S.; Houwing, A.F.P. The effect of fuel injection location on supersonic hydrogen combustion in a cavity-based model scramjet combustor. *Energies* **2020**, *13*, 193. [CrossRef]
26. Fotia, M.L.; Driscoll, J.F. Ram-Scram Transition and Flame/Shock-Train Interactions in a Model Scramjet Experiment. *J. Propuls. Power* **2013**, *29*, 261–273. [CrossRef]
27. Rasmussen, C.C.; Dhanuka, S.K.; Driscoll, J.F. Visualization of flameholding mechanisms in a supersonic combustor using PLIF. *Proc. Combust. Inst.* **2007**, *31*, 2505–2512. [CrossRef]
28. Xi, W.X.; Wang, Z.G.; Li, Q.; Liang, J.H. Forced ignition of kerosene spray in supersonic flowfield with hot gas injection. *J. Aerosp. Power* **2012**, *27*, 2436–2441.
29. Xi, W.X.; Wang, Z.G.; Sun, M.B.; Liu, W.D.; Li, Q. Experimental investigation of ignition transient phase in model supersonic combustor. *J. Aerosp. Eng.* **2014**, *27*, 04014009.

MDPI

Article

Improved Body Force Model for Estimating Off-Design Axial Compressor Performance

Jia Huang [1], Yongzhao Lv [2,3,*], Aiguo Xia [2], Shengliang Zhang [2], Wei Tuo [2], Hongtao Xue [2], Yantao Sun [2] and Xiuran He [2]

1 School of Aeronautics and Astronautics, Central South University, Changsha 410083, China; huangjia216@126.com or huangjia2019@csu.edu.cn
2 Beijing Aeronautical Engineering Technical Research Center, Beijing 100076, China; 18601251726@163.com (A.X.); zsl5046@sina.com (S.Z.); tirwit@163.com (W.T.); hunter_share1972@sina.com (H.X.); yts@buaa.edu.cn (Y.S.); hexiuran@163.com (X.H.)
3 Unit 93609 of the Chinese People's Liberation Army, Hohhot 010100, China
* Correspondence: lvjiuhui@163.com; Tel.: +86-15-810-534-934

Abstract: Based on the COMSOL software, body forces substituted into the Reynolds-averaged Navier–Stokes (RANS) equations as the source terms instead of the actual blade rows were improved to better predict the compressor performance. Improvements in parallel body force modeling were implemented, central to which were the local flow quantities. This ensured accurate and reliable off-design performance prediction. The parallel force magnitude mainly depended on the meridional entropy gradient extracted from three-dimensional (3D) steady single-passage RANS solutions. The COMSOL software could easily and accurately translate the pitchwise-averaged entropy into the grid points of the body force domain. A NASA Rotor 37 was used to quantify the improved body force model to represent the compressor. Compared with the previous model, the improved body force model was more efficient for the numerical calculations, and it agreed well with the experimental data and computational fluid dynamics (CFD) results. The results indicate that the improved body force model could quickly and efficiently capture the flow field through a turbomachinery blade row.

Keywords: improved body force model; off-design; Reynolds-averaged Navier–Stokes equations; meridional entropy gradient; rotor 37

Citation: Huang, J.; Lv, Y.; Xia, A.; Zhang, S.; Tuo, W.; Xue, H.; Sun, Y.; He, X. Improved Body Force Model for Estimating Off-Design Axial Compressor Performance. *Energies* **2022**, *15*, 4389. https://doi.org/10.3390/en15124389

Academic Editor: Adonios Karpetis

Received: 29 March 2022
Accepted: 3 June 2022
Published: 16 June 2022

1. Introduction

Because axial compressors often operate under off-design conditions, aircraft engines must have good performance [1]. The performance of the compressor under off-design conditions must be accurately estimated during the engine design stage [2,3]. Computational fluid dynamics (CFDs) have been considerably developed in recent decades and can accurately and efficiently simulate the flow field and estimate the axial compressor performance [4–6]. However, 3D, unsteady, multi-row calculations through the compressor or a coupled inlet fan with the actual blade geometry demand significant computer resources, including CPU time and memory [7,8]. Moreover, lots of flow field simulations are required [9,10]. To save computer resources, a passage-averaged body force model substituted into the RANS equations as the source terms instead of the actual blade rows is proposed herein and used to simulate the pressure rise, flow turning, and loss effects caused by the rotor/stator blade rows with reasonable computer resources [11,12].

Kim developed a body force approach to simulate the flow fields of N3-X. It was assumed that entropy production was related to the mass flow rate (MFR) at a constant speed [13,14]. Li et al. developed a method to perform the coupled inlet-fan Navier–Stokes simulation using the COMSOL–CFD code [15]. The advantage of this method was that the COMSOL–CFD code is a completely open architecture, and the flux terms in the RANS equations can be altered without compromising the computational stability. Source terms and boundary conditions can be functions or logical expressions of arbitrary variables. The body force terms were inputted into the COMSOL–CFD code and could directly simulate the viscous flow close to the blade passage walls and the momentum exchange between fluids. However, the parallel force model had limitations when calculating the local parallel force magnitude, which was proportional to the square of the total relative velocity [13]. The adjusting function $g(\dot{m}_{local})$ in the normal force formula, which depends on the local MFR, also makes the numerical simulation complicated and tedious [14,15]. Based on the work [15], an improved body force model was proposed using the COMSOL–CFD code. The COMSOL software can easily facilitate all steps in the modeling, part definition, meshing, simulation, and data post processing processes. In this study, the parallel force formula was modified using Marble's results [16], which indicated that the parallel force magnitude was proportional to the meridional entropy gradient. The magnitude of the meridional entropy gradient was estimated from the 3D steady single-passage RANS solutions of the compressor. The COMSOL software could accurately translate the pitchwise-averaged entropy into the grid points of the body force domain. The normal force formula no longer contained the term $g(\dot{m}_{local})$, which made the simulation processes efficient. A NASA Rotor 37 was used to quantify the improved body force model to represent the compressor.

The rest of this paper is organized as follows: Section 2 introduces the construction of the improved model; Section 3 explains the computational case and numerical techniques for flow simulations; Section 4 verifies this model, and in closing, Section 5 presents an overall conclusion drawn.

2. Improved Body Force Model

2.1. Governing Equation

In this paper, the governing equations [15] in Cartesian coordinates can be written in a non-conservative form, and the body force terms were added on the right-hand side as source terms, as follows.

$$\begin{array}{l} b\left[\frac{\partial\rho}{\partial t}+\nabla\cdot(\rho V)\right] \\ b\left[\rho\frac{DV_x}{Dt}+\frac{\partial p}{\partial x}-\frac{\partial\tau_{xx}}{\partial x}-\frac{\partial\tau_{yx}}{\partial y}-\frac{\partial\tau_{zx}}{\partial z}\right] \\ b\left[\rho\frac{DV_y}{Dt}+\frac{\partial p}{\partial y}-\frac{\partial\tau_{xy}}{\partial x}-\frac{\partial\tau_{yy}}{\partial y}-\frac{\partial\tau_{zy}}{\partial z}\right] \\ b\left[\rho\frac{DV_z}{Dt}+\frac{\partial p}{\partial z}-\frac{\partial\tau_{xz}}{\partial x}-\frac{\partial\tau_{yz}}{\partial y}-\frac{\partial\tau_{zz}}{\partial z}\right] \\ b\left[\begin{array}{l} \rho\frac{De}{Dt}-(\frac{\partial}{\partial x}\left(k\frac{\partial T}{\partial x}\right)+\frac{\partial}{\partial y}\left(k\frac{\partial T}{\partial y}\right)+\frac{\partial}{\partial z}\left(k\frac{\partial T}{\partial z}\right)-\frac{\partial(up)}{\partial x}-\frac{\partial(vp)}{\partial y}-\frac{\partial(wp)}{\partial z}+\frac{\partial(u\tau_{xx})}{\partial x} \\ +\frac{\partial(u\tau_{yx})}{\partial y}+\frac{\partial(u\tau_{zx})}{\partial z}+\frac{\partial(v\tau_{xy})}{\partial x}+\frac{\partial(v\tau_{yy})}{\partial y}+\frac{\partial(u\tau_{zy})}{\partial z}+\frac{\partial(w\tau_{xz})}{\partial x}+\frac{\partial(w\tau_{yz})}{\partial y}+\frac{\partial(w\tau_{zz})}{\partial z}) \end{array}\right] \end{array} = \Phi \quad (1)$$

where $\tau_{xx} = \frac{2}{3}\mu\left(2\frac{\partial u}{\partial x}-\frac{\partial v}{\partial y}-\frac{\partial w}{\partial z}\right)$, $\tau_{xy}=\tau_{yx}=\mu\left(\frac{\partial v}{\partial x}+\frac{\partial u}{\partial y}\right)$, $\tau_{yy}=\frac{2}{3}\mu\left(-\frac{\partial u}{\partial x}+2\frac{\partial v}{\partial y}-\frac{\partial w}{\partial z}\right)$, $\tau_{xz}=\tau_{zx}=\mu\left(\frac{\partial u}{\partial z}+\frac{\partial w}{\partial x}\right)$, $\tau_{zz}=\frac{2}{3}\mu\left(-\frac{\partial u}{\partial x}-\frac{\partial v}{\partial y}+2\frac{\partial w}{\partial z}\right)$, $\tau_{yz}=\tau_{zy}=\mu\left(\frac{\partial w}{\partial y}+\frac{\partial v}{\partial z}\right)$, the energy e is given by $e = C_vT+0.5\left(V_x^2+V_y^2+V_z^2\right)$; the pressure p is given by $p = (\gamma-1)\left(e-0.5\rho\left(V_x^2+V_y^2+V_z^2\right)\right)$; the blockage b can be modeled as $b=\left|\theta_s-\theta_p\right|N/2\pi$ to account for the blade thickness, and the source term Φ is given by $\Phi = \Phi' + \Phi''$. Φ' and Φ'' are body forces and extra terms, respectively, as follows.

$$\Phi' = b\left(0,\rho f_x,\rho f_y,\rho f_z,\rho\left(V_xf_x+V_yf_y+V_zf_z\right)\right)^T \quad (2)$$

$$\Phi'' = \left\{ \begin{array}{l} b(\frac{\partial \rho V_\theta}{\partial \theta} - r\Omega \frac{\partial \rho}{\partial \theta}) - \rho V \cdot \nabla b \\ b(\frac{\partial(\rho V_x V_\theta - \tau_{x\theta})}{\partial \theta} - r\Omega \frac{\partial \rho V_x}{\partial \theta}) - bV_z(\frac{\partial \rho V_\theta}{\partial \theta} - r\Omega \frac{\partial \rho}{\partial \theta}) \\ b(\frac{\partial(\rho V_\theta V_r - \tau_{r\theta})}{\partial \theta} - r\Omega \frac{\partial \rho V_r}{\partial \theta} - V_r(\frac{\partial \rho V_\theta}{\partial \theta} - r\Omega \frac{\partial \rho}{\partial \theta}))\cos\theta - b(\frac{\partial(\rho V_\theta^2 + p - \tau_{\theta\theta})}{\partial \theta} - r\Omega \frac{\partial \rho V_\theta}{\partial \theta} - V_\theta(\frac{\partial \rho V_\theta}{\partial \theta} - r\Omega \frac{\partial \rho}{\partial \theta}))\sin\theta \\ b(\frac{\partial(\rho V_\theta V_r - \tau_{r\theta})}{\partial \theta} - r\Omega \frac{\partial \rho V_r}{\partial \theta} - V_r(\frac{\partial \rho V_\theta}{\partial \theta} - r\Omega \frac{\partial \rho}{\partial \theta}))\sin\theta + b(\frac{\partial(\rho V_\theta^2 + p - \tau_{\theta\theta})}{\partial \theta} - r\Omega \frac{\partial \rho V_\theta}{\partial \theta} - V_\theta(\frac{\partial \rho V_\theta}{\partial \theta} - r\Omega \frac{\partial \rho}{\partial \theta}))\cos\theta \\ b(\frac{\partial(V_\theta(e+p) - V_x\tau_{x\theta} - V_\theta\tau_{\theta\theta} - V_r\tau_{r\theta} + \dot{q}_\theta)}{\partial \theta} - r\Omega \frac{\partial e}{\partial \theta}) - e\left[b(\frac{\partial \rho V_\theta}{\partial \theta} - r\Omega \frac{\partial \rho}{\partial \theta}) - \rho V \cdot \nabla b\right] + (V_x(e+p) - V_x\tau_{xx} - V_y\tau_{xy} - V_z\tau_{xz} + \dot{q}_x)\frac{\partial b}{\partial x} \\ +(V_y(e+p) - V_x\tau_{yx} - V_y\tau_{yy} - V_z\tau_{yz} + \dot{q}_y)\frac{\partial b}{\partial y} + (V_z(e+p) - V_x\tau_{zx} - V_y\tau_{zy} - V_z\tau_{zz} + \dot{q}_z)\frac{\partial b}{\partial z} \end{array} \right\} \quad (3)$$

In Equation (1), the energy equation of the fluid contains the internal energy and the mechanical energy. So, the differential form of the energy equation can be written as

$$\rho \frac{D}{Dt}\left(E + \frac{1}{2}u_i u_i\right) = \frac{\partial}{\partial x_i}(u_j \sigma_{ij}) - \frac{\partial q_i}{\partial x_i} + \rho u_i f_i \quad (4)$$

and the differential form of the momentum equation can be written as

$$\rho \frac{Du_i}{Dt} = \frac{\partial \sigma_{ij}}{\partial x_j} + \rho f_i \quad (5)$$

Then, by minusing $u_i \times$ Equation (5), the energy equation becomes:

$$\rho \frac{DE}{Dt} = \sigma_{ij} \frac{\partial u_j}{\partial x_i} - \frac{\partial q_i}{\partial x_i} \quad (6)$$

From Equation (6), it can be seen that the body force can only change the size of the mechanical energy with nothing on the internal energy. So, Equation (1) can be written as

$$\begin{array}{l} b\left[\frac{\partial \rho}{\partial t} + \nabla \cdot (\rho V)\right] \\ b\left[\rho \frac{DV_x}{Dt} + \frac{\partial p}{\partial x} - \frac{\partial \tau_{xx}}{\partial x} - \frac{\partial \tau_{yx}}{\partial y} - \frac{\partial \tau_{zx}}{\partial z}\right] \\ b\left[\rho \frac{DV_y}{Dt} + \frac{\partial p}{\partial y} - \frac{\partial \tau_{xy}}{\partial x} - \frac{\partial \tau_{yy}}{\partial y} - \frac{\partial \tau_{zy}}{\partial z}\right] \\ b\left[\rho \frac{DV_z}{Dt} + \frac{\partial p}{\partial z} - \frac{\partial \tau_{xz}}{\partial x} - \frac{\partial \tau_{yz}}{\partial y} - \frac{\partial \tau_{zz}}{\partial z}\right] \\ b\left[\begin{array}{l} \rho \frac{DE}{Dt} + p\nabla \cdot V - \frac{\partial}{\partial x}\left(k\frac{\partial T}{\partial x}\right) - \frac{\partial}{\partial y}\left(k\frac{\partial T}{\partial y}\right) - \frac{\partial}{\partial z}\left(k\frac{\partial T}{\partial z}\right) - \tau_{xx}\frac{\partial V_x}{\partial x} - \tau_{xy}\frac{\partial V_y}{\partial x} - \tau_{xz}\frac{\partial V_z}{\partial x} \\ -\tau_{yx}\frac{\partial V_x}{\partial y} - \tau_{yy}\frac{\partial V_y}{\partial y} - \tau_{yz}\frac{\partial V_z}{\partial y} - \tau_{zx}\frac{\partial V_x}{\partial z} - \tau_{zy}\frac{\partial V_y}{\partial z} - \tau_{zz}\frac{\partial V_z}{\partial z} \end{array}\right] \end{array} = \Phi + \Phi''' \quad (7)$$

where

$$\Phi''' = (0\,0\,0\,0 \; - \; (V_x(\Phi'_x + \Phi''_x) + V_y(\Phi'_y + \Phi''_y) + V_z(\Phi'_z + \Phi''_z)))^T \quad (8)$$

2.2. Construction of the Body Force

In this model, the blade force on a cascade section was separated into two parts that were parallel and normal to the local flow, as shown in Figure 1. The normal force F_n represents the effects of the pressure difference, and the parallel force F_p is related to the viscous shear.

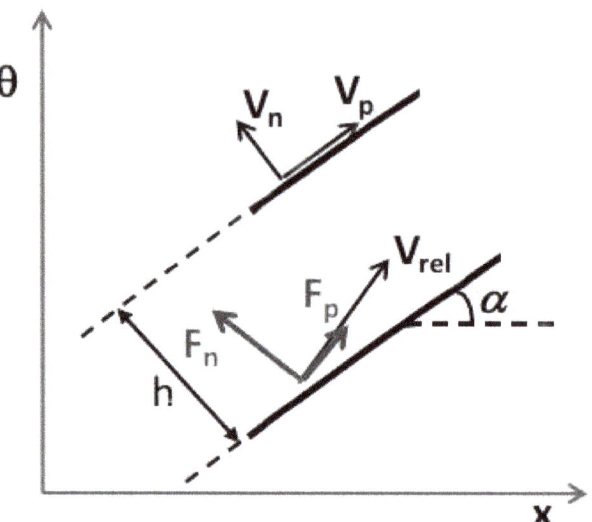

Figure 1. Illustration of body forces parallel and normal to the local flow.

2.2.1. Normal Body Force

The body force normal to the flow direction, F_n, is formulated as follows:

$$F_n = \frac{K_n}{h} V_n V_p + \frac{2}{c}\sin(\frac{\Delta\alpha}{2}) V_n^2 \quad (9)$$

where h, V_n, V_p, and $\Delta\alpha$ are the blade-to-blade gap-staggered spacing, axial velocity, circumferential velocity, and the camber angle difference between the trailing edge and leading edge, respectively. K_n is the normal force coefficient [13] formulated in Equation (10). The second expression on the right-hand side of Equation (9) was different from Gong's formulation:

$$K_n = (4.2 - 3.3\alpha) f(r) \quad (10)$$

Here, the second expression *f(r)* was used to adjust the normal force coefficient K_n, which was a line segment connected by a few control points along the spanwise direction. The components of the normal force F_n in the x, y, and z directions are expressed as:

$$\begin{aligned} F_{n,x} &= F_n \frac{V_z \cos\theta - V_y \sin\theta}{V_{rel}} \\ F_{n,y} &= -F_n \frac{V_x}{V_{rel}} \sin\theta, \\ F_{n,z} &= F_n \frac{V_x}{V_{rel}} \cos\theta \end{aligned} \quad (11)$$

where V_{rel} is the relative blade velocity.

2.2.2. Parallel Body Force

The parallel body force is always tangential to the relative flow and represents the effects of the mixing of the tip leakage flow and main flow and flow blade surface boundary layers. The parallel body force F_p is formulated as follows [11]:

$$F_p = -\frac{K_p}{h} V_{rel}^2 \quad (12)$$

where the parallel force coefficient K_p is equal to 0.04. In a fixed speed, the magnitude of F_p increases as the MFR increases. The entropy production across the blade domain decreases as the MFR increases. The parallel body force F_p can be expressed as:

$$F_p = -T\frac{V_m}{V_{rel}}\frac{\partial s}{\partial m} \tag{13}$$

where s is the entropy; T is the temperature, and m is the coordinate along the meridional streamline (Figure 2). Instead of Gong's model, Equation (13) was used herein to calculate the magnitude of the local parallel force because the direct relationship between the meridional entropy gradient and the parallel body force makes the modeling process of the body force easier and more physical.

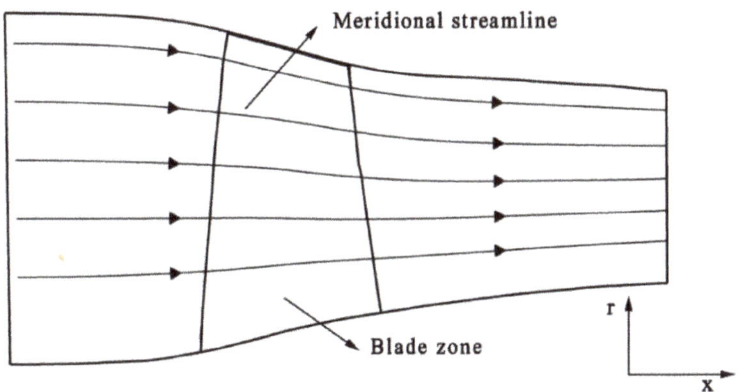

Figure 2. Schematic diagram of the meridional streamline.

The velocity V_m and the partial differential ∂_m can be written using the follow -ing transformations:

$$\begin{aligned} dm^2 &= dx^2 + dr^2 \\ V_m^2 &= V_x{}^2 + V_r{}^2 \\ V_m\partial_m &= V_x\partial_x + V_r\partial_r \end{aligned} \tag{14}$$

Hence, Equation (13) becomes simply

$$F_p = -\frac{T}{V_{rel}}(V_x\frac{\partial s}{\partial x} + V_r\frac{\partial s}{\partial r}) \tag{15}$$

where ∂_s/∂_x and ∂_s/∂_r are the local entropy gradients along the axial and radial directions, respectively. The local gradients of the pitchwise-averaged entropy generated across the blade rows extracted from the 3D steady single-passage RANS solutions are used as the input terms to Equation (15). The components of the parallel force F_p in the x, y, and z directions are expressed as:

$$F_{p,x} = F_p\frac{V_x}{V_{rel}},\ F_{p,y} = F_p\frac{V_y}{V_{rel}},\ F_{p,z} = F_p\frac{V_z}{V_{rel}} \tag{16}$$

3. Computational Case and Numerical Techniques

Figure 3 briefly shows the demand on the computational accuracy versus the computer resources for different levels of numerical techniques [17]. It is obvious that the improved body force model can be relatively independent of the empiricism and cost much less computer resources than the RANS model within a proper accuracy range [18]. Below, the RANS model and improved body force model were used to simulate the flow field performances of Rotor 37. Compared with the experimental data, the results were analyzed to further validate the accuracy of the improved body force model.

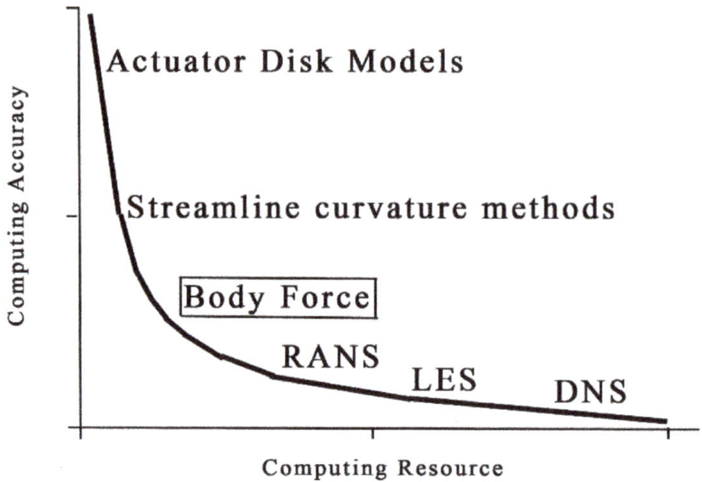

Figure 3. Demands of models on accuracy versus resources.

3.1. Compressor Used for Study

A transonic compressor, a NASA Rotor 37, as shown in Figure 4, was used to validate the improved body force model because it is a well-documented and typical test case. This section presents the CFD tool and methodologies. The main design parameters are summarized in Table 1.

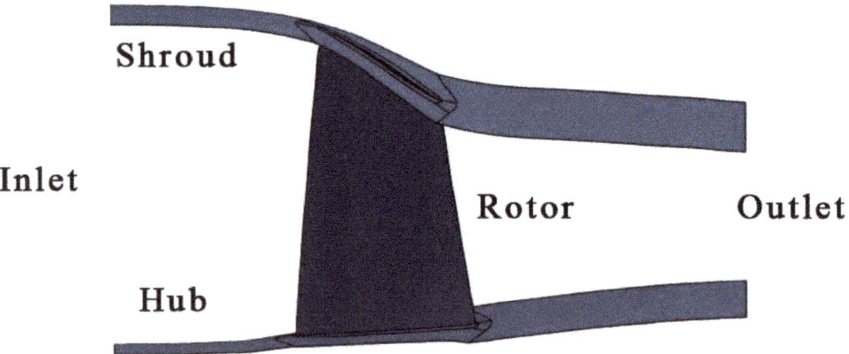

Figure 4. Schematic of the NASA Rotor 37.

Table 1. Design parameters of the NASA Rotor 37.

Parameters	Value
Blade number	36
Inlet hub-to-tip ratio	0.7
Blade aspect ratio	1.19
Tip solidity	1.29
Tip relative inlet Ma	1.48
Rotating speed (rpm)	17,188

Table 1. *Cont.*

Parameters	Value
Mass flow rate (kg/s)	20.19
Total pressure ratio	2.106
Adiabatic efficiency (%)	87.7

3.2. CFD Methods

3.2.1. Turbomachinery Flow Simulation

In this study, the commercial solver NUMECA FINE was used as the CFD tool for 3D steady single-passage flow simulations of the compressor. And then later on, the meridional entropy gradient of the compressor was extracted from steady RANS solutions simulated by the NUMECA FINE to calculate the parallel force magnitude. Finally, the experimental data and NUMECA FINE results were used to compare with results obtained by the improved body force model. In detail, the temporal and spatial discretization schemes were selected as the explicit fourth-order Runge–Kutta scheme and second-order accurate central difference scheme, respectively. Based on previous studies [15,19,20], the one equation Spalart–Allmaras (S–A) turbulence model was used. Some acceleration techniques, such as implicit residual smoothing and local time stepping methods, were employed [21]. The single-blade passage simulation was performed with periodic boundary conditions in the circumferential direction. At the inlet, the total temperature and pressure were specified along with the flow angle. The outlet of the computational domain was located at approximately two chords downstream of the rotor. At the outlet, based on the radial equilibrium, the averaged static pressure was given. Adiabatic and no-slip conditions were given on solid surfaces.

In this paper, all computations were performed using identical boundary conditions. Figure 5 shows the computational meshes for the Rotor 37. A periodic multi-block O4H-type structured grid was used in each blade channel. An O-type grid and an H-type grid were employed around the blade surface and the remaining regions, respectively. The minimum grid orthogonal angle was greater than 30°, and y^+ near the wall was less than 5, as shown in Figure 6.

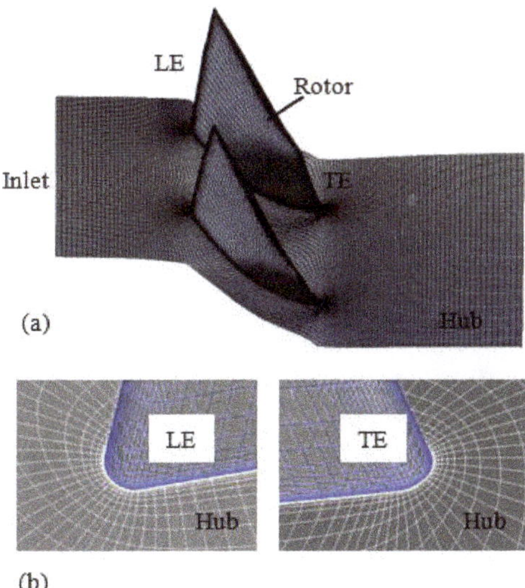

Figure 5. NUMECA FINE computational mesh. (**a**) Perspective view and (**b**) View from casing at LE and TE of hub.

A series of computations were conducted with four different meshes to verify the solution errors related to the grid by imposing the same boundary conditions, and the MFR and adiabatic efficiency are shown in Figure 7. The figure illustrates that the adiabatic efficiency and MFR remained basically the same when the mesh number reached 734,761. So, the grid that consists of a total of 734,761 meshes was selected to achieve the mesh independence needed to provide the flow field analysis in detail.

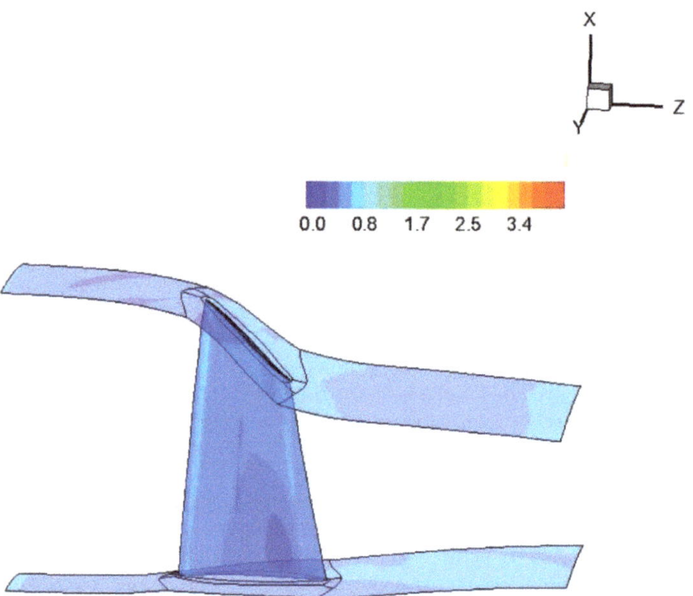

Figure 6. Distribution of y+ on walls of the hub, blade, and shroud.

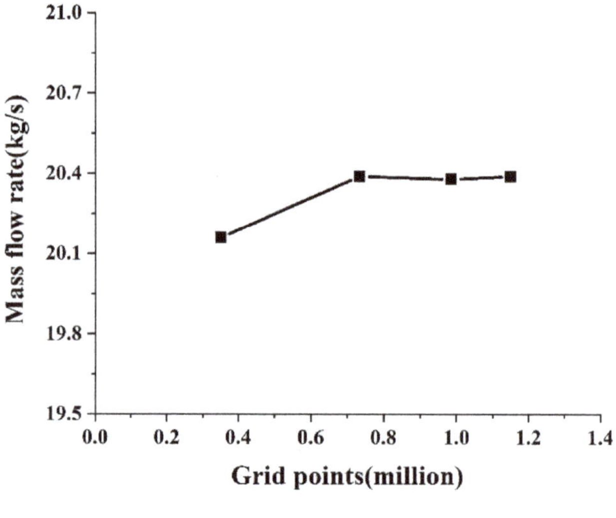

(a)

Figure 7. *Cont.*

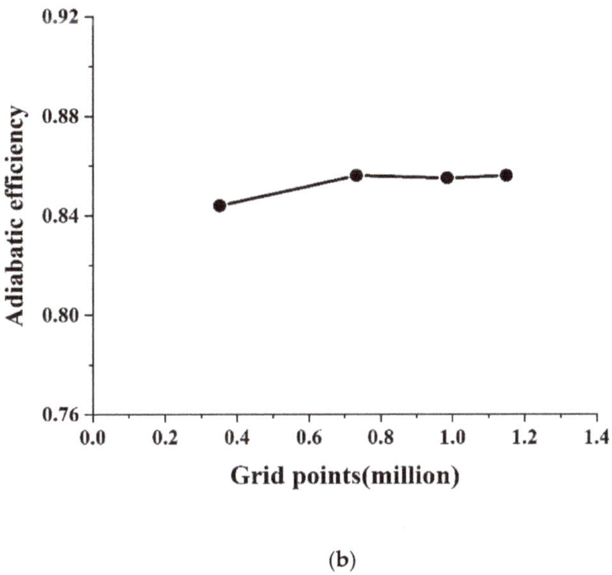

(**b**)

Figure 7. Mesh independence verification. (**a**) Mass flow rate and (**b**) Adiabatic efficiency.

3.2.2. HMNF Flow Simulation

High Mach Number Flow (HMNF) as a COMSOL–CFD module was selected to add body forces into the governing equations for simulating the pressure rise, flow turning, and loss effects caused by the blade rows in this study. The finite element method was employed to discretize the RANS equations, and the one equation S–A turbulence model was used the same as the turbomachinery flow simulation. In the HMNF module, segregated solvers were used to compute the flux, in which Newton's method was executed. For 3D numerical full-annulus simulations, the full-annulus grid of the compressor was chosen as a hexahedral structure grid, and local encryption near the wall was carried out. The boundary conditions were the same as in the NUMECA FINE simulations. The HMNF module computational mesh for the Rotor 37 is shown in Figure 8, and the rotor region is marked in blue. The grid consists of a total of 75,600 meshes and is appropriate in achieving the mesh independence needed to provide the flow field analysis in detail, as shown in Figure 9.

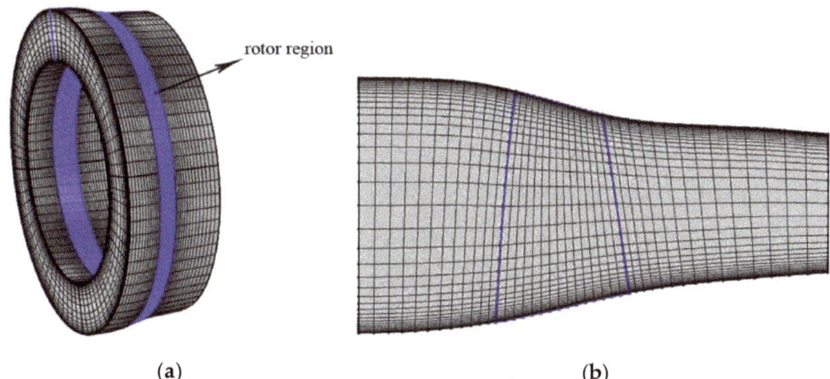

(**a**) (**b**)

Figure 8. HMNF module computational mesh. (**a**) 3D view and (**b**) Meridional plane view.

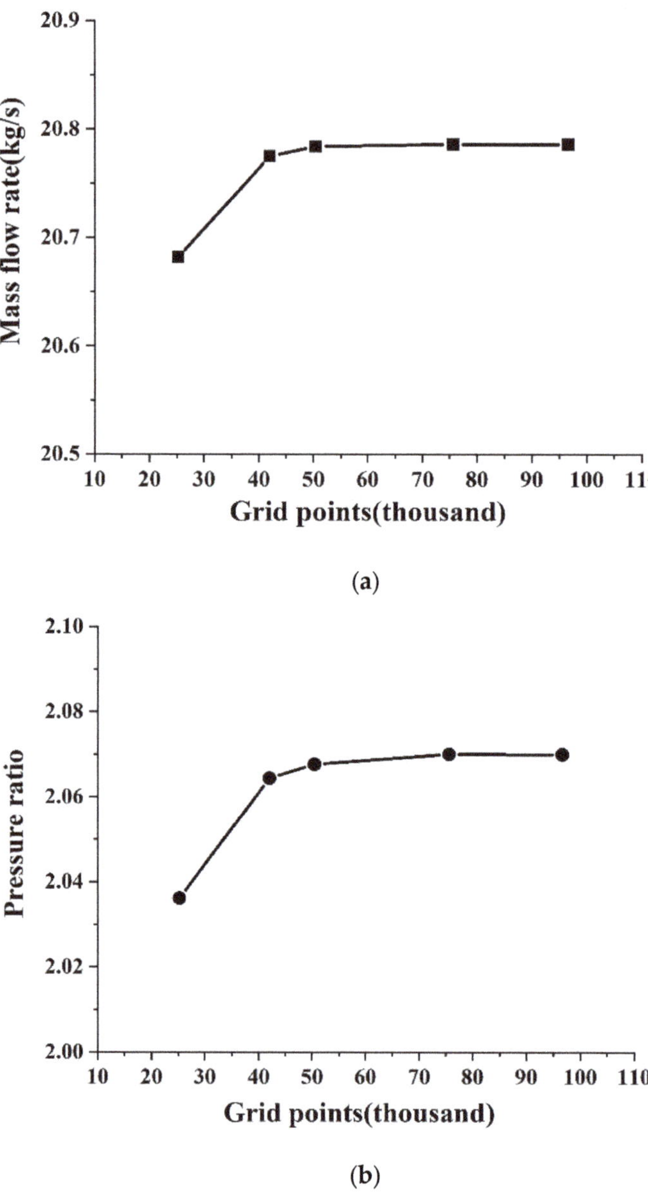

Figure 9. Mesh independence verification. (**a**) Mass flow rate; (**b**) Pressure ratio.

According to the above formulas, the magnitude of the body force is mainly determined by the local flow field and blade geometry parameters. Before the numerical calculations, it is necessary to discretize the blade geometry parameters into the body force domain grid points, and it mainly includes the camber angle α, the blade-to-blade gap-staggered spacing h, the blockage b, and the solidity σ as shown in Figure 10. The local gradient of the pitchwise-averaged entropy as an input term to the parallel force formula was obtained from steady RANS solutions simulated by the commercial solver NUMECA FINE in this study. Figure 11 shows the pitchwise-averaged entropy on the meridional plane for a 98% choked mass flow at 100% of the designed rotor speed. The number in Figure 11 represents the value of the isentropic curve. The values of the entropy around the hub and shroud were larger, and the loss, which degraded the compressor performance, was also larger. Because the body forces were added into the governing equations as source terms, the improved body force formula was defined at grid points of the body force domain. However, the body force model grid is different from the pitchwise-averaged NUMECA FINE grid. The COMSOL software could easily and accurately translate the body force formula inputs extracted from solutions solved by the NUMECA FINE to grid points of the body force domain. Then, the values of the entropy gradient along the axial and radial directions at those grid points could be determined.

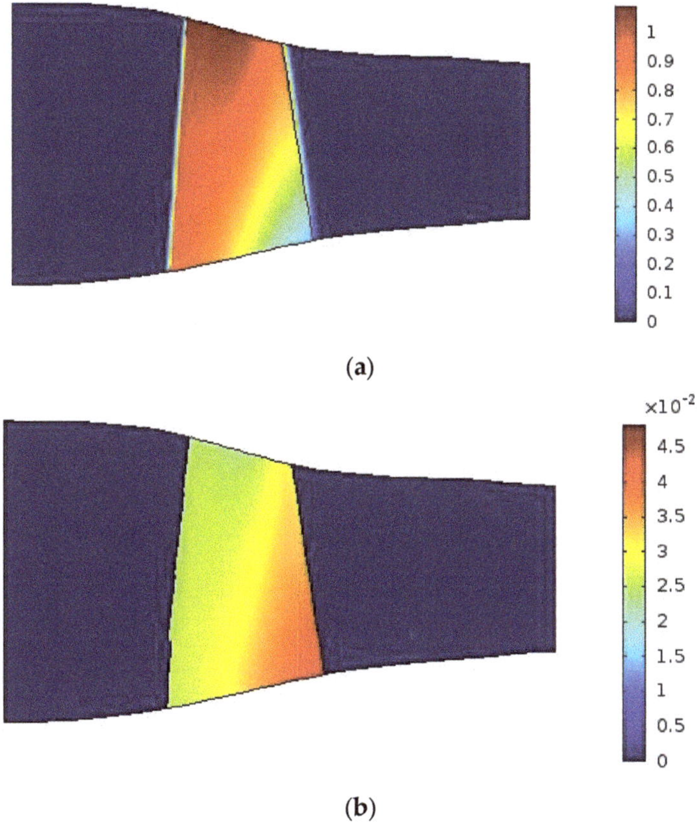

Figure 10. *Cont.*

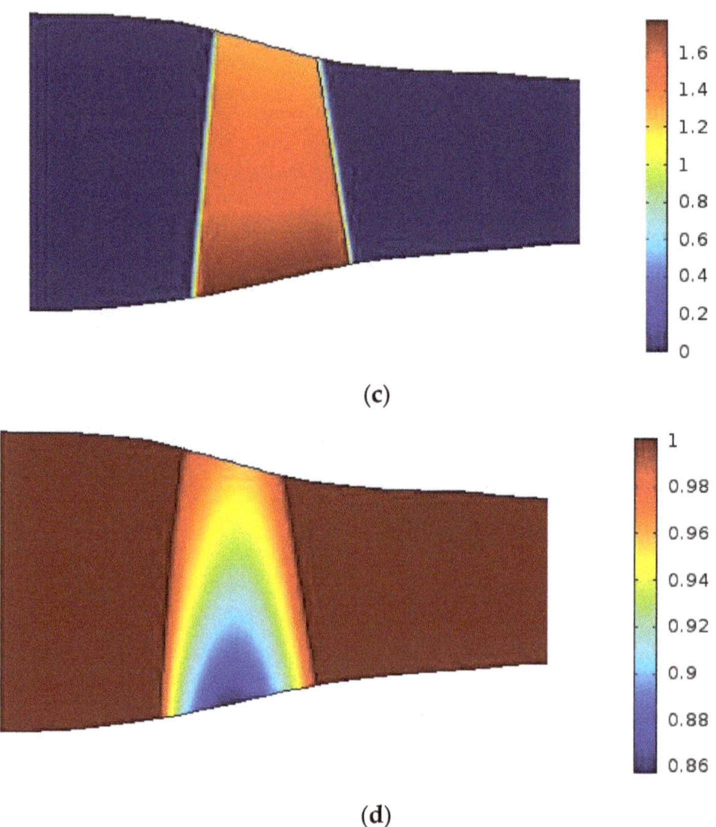

Figure 10. NASA Rotor 37 blade geometry parameters. (**a**) The camber angle α (rad); (**b**) The spacing h (m); (**c**) The solidity σ (1); and (**d**) The blockage b (1).

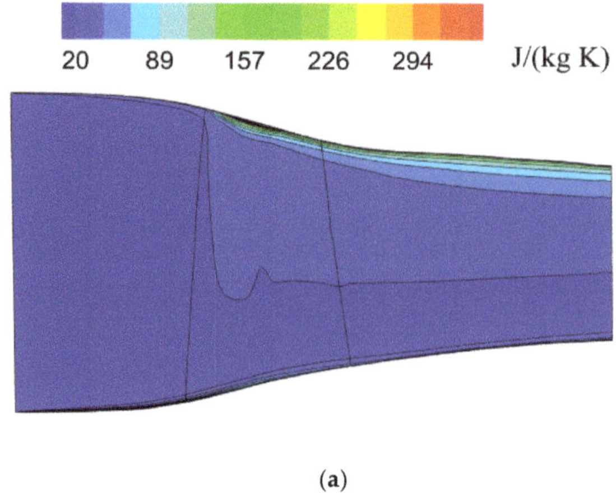

(**a**)

Figure 11. *Cont.*

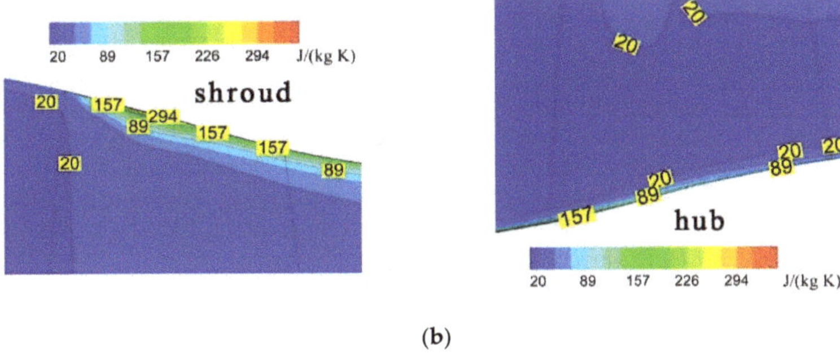

(b)

Figure 11. The pitchwise-averaged entropy contours. (**a**) View on the meridional plane and (**b**) View around shroud and hub.

4. Model Validation

The flow field performances of the Rotor 37 were simulated using the NUMECA FINE, and they improved the body force model. The simulated results were analyzed to further validate the accuracy of the improved body force model.

Figure 12 illustrates a comparison of the Rotor 37 pressure ratio versus the MFR at 80%, 90%, and 100% of the designed rotor speed. It was shown that NUMECA FINE and the improved body force model results agreed very well with the experimental data obtained from the AGARD Advisory Report 355 entitled *CFD Validation for Propulsion System Components*. The experimental values were slightly higher than the simulation results. The maximum error was 1.2%. At other speeds, the simulation that used the improved body force model generally agreed well with the NUMECA results. The maximum error was 1.9%, which was still within the acceptable error range. Compared with Li's model [15], the results using the improved body force model were better at lower MFR points. Therefore, the improved body force model could capture the flow field through a turbomachinery blade row well within a proper accuracy range.

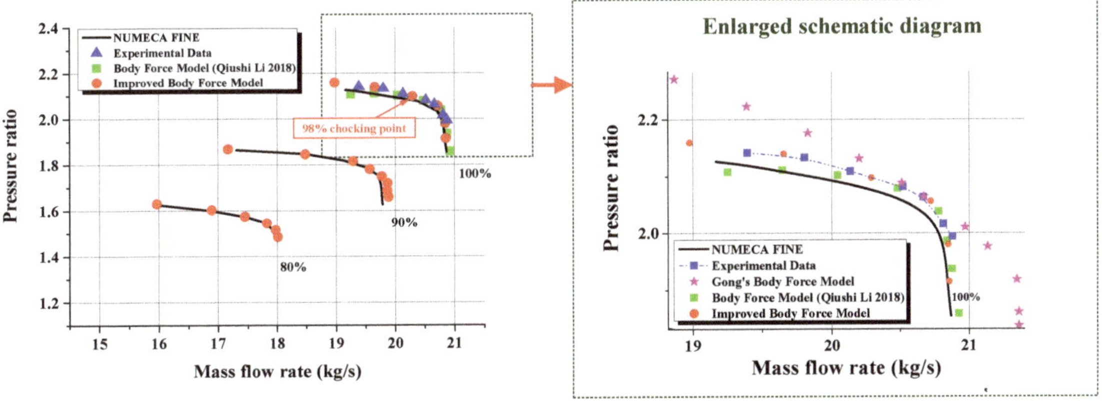

Figure 12. Rotor 37 pressure ratio versus mass flow rate.

Figure 13 presents the pitch-averaged total temperature ratio, total pressure ratio, Mach number, and swirl angle at the outlet along the spanwise direction for a 98% choked mass flow (Figure 12) for the experimental data, NUMECA FINE, and body force models. Compared with the Gong and Li models, the distributions of the performance parameters around the hub and shroud obtained using the improved body force model were better and agreed well with the experimental data. The errors about the NUMECA FINE and improved body force model in the comparisons with the experimental data in Figure 13a–c are listed in Table 2. Figure 14 presents the Mach number contours, pressure contours, total pressure contours, and swirl angle contours on the meridional plane for a 98% choked mass flow. The comparison between the NUMECA FINE and the improved body force model indicated that the results' overall distribution trends were basically the same. In the rotor region illustrated in Figure 14, it could be approximated that the pressure and total pressure increased linearly along the streamwise direction. The inlet airflow was also deflected through the blade passage, with the flow angle increasing almost linearly.

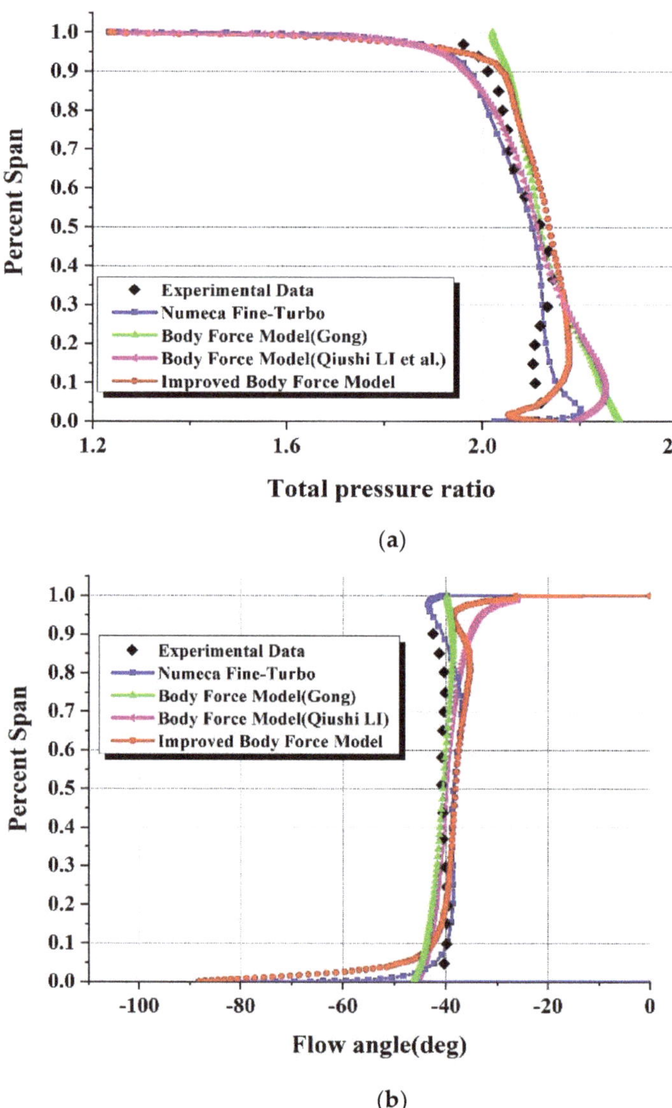

Figure 13. *Cont.*

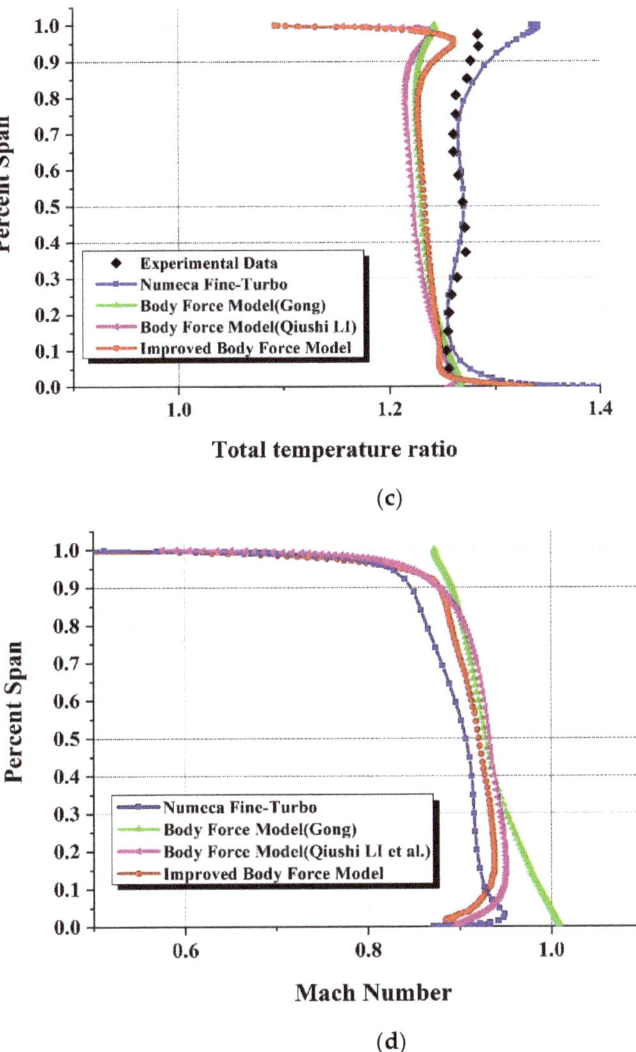

Figure 13. Comparison of Rotor 37 flow characteristics at outlet along the span. (**a**) Total pressure ratio distribution; (**b**) Swirl angle distribution; (**c**) Total temperature distribution; and (**d**) Mach distribution.

Table 2. The errors in the comparisons with the experimental data.

	NUMECA FINE	Improved Body Force Model
Figure 13a	3.2%	3.4%
Figure 13b	8.3%	14%
Figure 13c	3.7%	2.8%

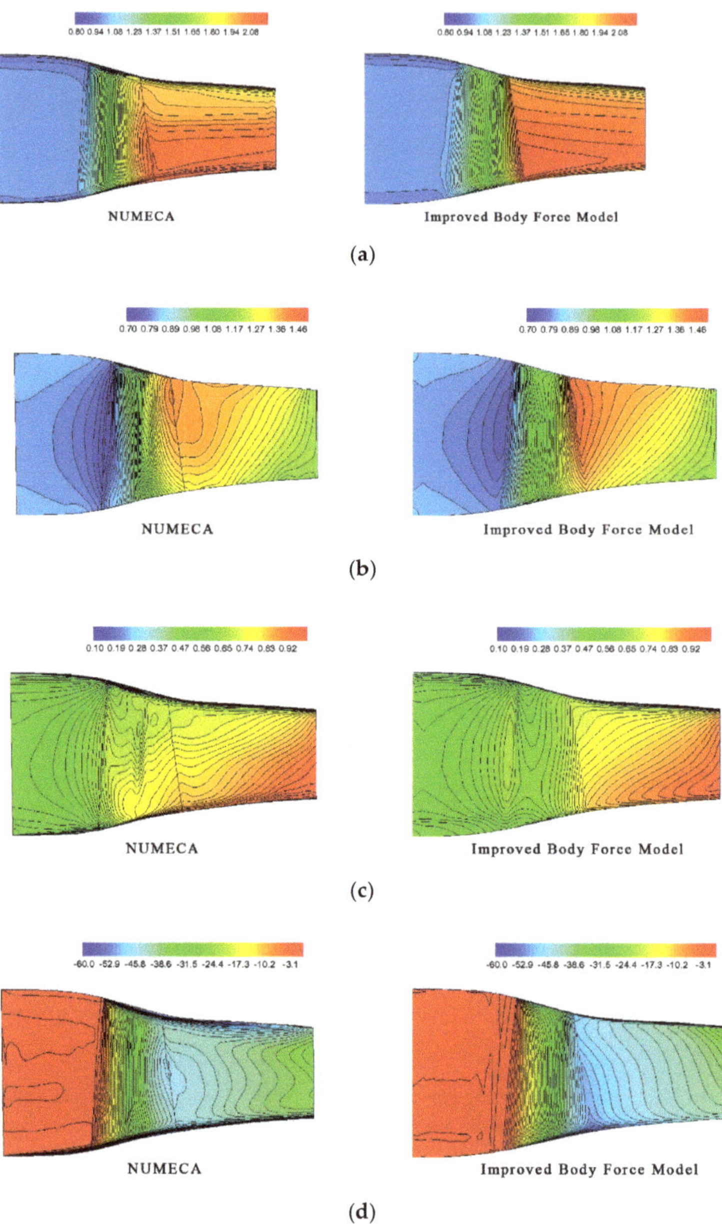

Figure 14. The pitchwise−averaged parameter contours on the meridional plane. (**a**) Dimensionless total pressure contours; (**b**) Dimensionless pressure contours; (**c**) Mach contours; and (**d**) Swirl angle contours.

It should be noted that based on the COMSOL software, this model could be further optimized in future work. It could be assumed that the local entropy gradients ∂_s/∂_x and ∂_s/∂_r are functions of the local parameter ρV_x, where V_x and ρ are the local axial velocity and density, respectively. The values of the local entropy gradients ∂_s/∂_x, ∂_s/∂_r and the local parameter ρV_x were extracted from the 3D steady single-passage RANS solutions. In that case, the flow field of aircraft/engine integration under clean and distorted inflows can be simulated easily using this model.

5. Conclusions

Body forces substituted into the RANS equations as the source terms instead of the actual blade rows solved using the COMSOL–CFD code were improved to better predict the compressor performance. The flow field performances of Rotor 37, including the pressure rise, flow turning, and loss effects caused by the blade rows were simulated by the improved body force model. Compared with the experimental data, NUMECA FINE, Gong's model, and Li's model the following conclusions can be drawn:

1. Based on the COMSOL software, the improved body force model could directly simulate the viscous flow close to the blade passage walls and the momentum exchange between fluids. The improved parallel force formula is modified using Marble's results, indicating that the magnitude of the parallel force is proportional to the entropy gradient along the meridional streamline extracted from 3D steady single-passage RANS solutions. Therefore, the modeling process of the body force is easier and more physical.
2. The improved normal force formula no longer contains the term $g(\dot{m}_{local})$, which makes the simulation processes more efficient. Compared with those of NUMECA FINE, the total number of grid points for the HMNF module is less at least an order of magnitude, and it takes less time to compute them. The simulation results for the Rotor 37 indicate that the improved body force model could capture the flow field through a turbomachinery blade row well within a proper accuracy range. When 3D, full-annulus, unsteady, multi-row calculations through the compressor or integrated calculations of the aircraft and engine were performed, the improved body forces could be substituted into the RANS equations as the source terms instead of the actual blade rows, and it could greatly reduce the total number of grid points and save significant computer resources, including CPU time and memory.
3. Because body forces are added into the governing equations as source terms, the improved body force formula was defined at grid points of the body force domain. The COMSOL software could easily and accurately translate the body force formula inputs extracted from solutions solved by the NUMECA FINE-Turbo EURANUS to grid points of the body force domain. The COMSOL software environment can easily facilitate all steps in the modeling, part definition, meshing, simulation, and data post processing processes.

Author Contributions: Conceptualization, Y.L.; validation, Y.L. and J.H.; investigation, Y.L., J.H. and A.X.; writing—original draft preparation, Y.L., S.Z. and W.T.; writing—review and editing, X.H. and Y.S.; supervision, H.X. All authors have read and agreed to the published version of the manuscript.

Funding: This research was funded by the Natural Science Foundation of Hunan Province, China (No. 2021JJ40741) and the National Science and Technology Major Project (No. J2019-V-0017-0012 and No. J2019-IV-0017-0085).

Institutional Review Board Statement: Not applicable.

Informed Consent Statement: Not applicable.

Data Availability Statement: Not applicable.

Conflicts of Interest: The authors declare no conflict of interest.

References

1. Ricci, M.; Mosele, S.G.; Benvenuto, M.; Astrua, P.; Pacciani, R.; Marconcini, M. Retrofittable Solutions Capability for Gas Turbine Compressors. *Int. J. Turbomach. Propuls. Power* **2022**, *7*, 3. [CrossRef]
2. Burberi, C.; Michelassi, V.; del Greco, A.S.; Lorusso, S.; Tapinassi, L.; Marconcini, M.; Pacciani, R. Validation of steady and unsteady CFD strategies in the design of axial compressors for gas turbine engines. *Aerosp. Sci. Technol.* **2020**, *107*, 106307. [CrossRef]

3. Li, Y.L.; Sayma, A.I. Computational fluid dynamics simulations of blade damage effect on the performance of a transonic axial compressor near stall. *Proc. Inst. Mech. Eng. Part C J. Mech. Eng. Sci.* **2015**, *229*, 2242–2260. [CrossRef]
4. Jian, H.; Hu, W. Numerical Investigation of Inlet Distortion on an Axial Flow Compressor Rotor with Circumferential Groove Casing Treatment. *Chin. J. Aeronaut.* **2008**, *21*, 496–505. [CrossRef]
5. Shen, X.B.; Wang, H.F.; Lin, G.P.; Bu, X.Q.; Wen, D.S. Unsteady simulation of aircraft electro-thermal deicing process with temperature-based method. *Proc. Inst. Mech. Eng. Part G J. Aerosp. Eng.* **2020**, *234*, 388–400. [CrossRef]
6. Sun, H.O.; Wang, M.; Wang, Z.Y.; Magagnato, F. Numerical investigation of surge prediction in a transonic axial compressor with a hybrid BDF/Harmonic Balance Method. *Aerosp. Sci. Technol.* **2019**, *90*, 401–409. [CrossRef]
7. Hu, Y.; Nie, C. Exploration of combined adjustment laws about IGV, stator and rotational speed in off-design conditions in an axial compressor. *Sci. China Technol. Sci.* **2010**, *53*, 969–975. [CrossRef]
8. Chu, F.; Dai, B.; Lu, N.; Ma, X.; Wang, F. Improved fast model migration method for centrifugal compressor based on bayesian algorithm and Gaussian process model. *Sci. China Technol. Sci.* **2018**, *61*, 1950–1958. [CrossRef]
9. Webster, R.; Sreenivas, K.; Hyams, D.; Hilbert, B.; Briley, W.; Whitfield, D. Demonstration of Sub-system Level Simulations: A Coupled Inlet and Turbofan Stage. In Proceedings of the 48th AIAA/ASME/SAE/ASEE Joint Propulsion Conference & Exhibit, Atlanta, GA, USA, 30 July–1 August 2012.
10. Li, S.; Liu, Y.; Omidi, M.; Zhang, C.; Li, H. Numerical Investigation of Transient Flow Characteristics in a Centrifugal Compressor Stage with Variable Inlet Guide Vanes at Low Mass Flow Rates. *Energies* **2021**, *14*, 7906. [CrossRef]
11. Gong, Y. A Computational Model for Rotating Stall Inception and Inlet Distortions in Multistage Compressors. Ph.D. Thesis, Massachusetts Institute of Technology, Cambridge, MA, USA, 1998.
12. Hsiao, E.; Naimi, M.; Lewis, J.P.; Dalbey, K.; Gong, Y.; Tan, C. Actuator duct model of turbomachinery components for powered-nacelle Navier-Stokes calculations. *J. Propuls. Power* **2001**, *17*, 919–927. [CrossRef]
13. Hyoungjin, K.; Mengsing, L. Flow simulation of N3X hybrid wing body configuration. In Proceedings of the 51st AIAA Aerospace Sciences Meeting including the New Horizons Forum and Aerospace Exposition, Grapevine, TX, USA, 7–10 January 2013.
14. Kim, H.; Liou, M.-S. Optimal Shape Design of Mail-Slot Nacelle on N3-X Hybrid Wing Body Configuration. In Proceedings of the 31st AIAA Applied Aerodynamics Conference, San Diego, CA, USA, 24–27 June 2013; p. 2413.
15. Qiushi, L.; Yongzhao, L.; Tianyu, P.; Da, L.; Ha'nan, L.; Yifang, G. Development of a coupled supersonic inlet-fan Navier–Stokes simulation method. *Chin. J. Aeronaut.* **2018**, *31*, 237–246.
16. Marble, F.E.; Hawthorne, W. Three-dimensional flow in turbomachines. *High Speed Aerodyn. Jet Propuls.* **1964**, *10*, 83–166.
17. Xu, L. Assessing viscous body forces for unsteady calculations. *J. Turbomach.* **2003**, *125*, 425–432. [CrossRef]
18. Ritos, K.; Kokkinakis, I.W.; Drikakis, D. Performance of High-Order Implicit Large Eddy Simulations. *Comput. Fluids* **2018**, *173*, 307–312. [CrossRef]
19. Lange, M.; Vogeler, K.; Mailach, R.; Gomez, S.E. An experimental verification of a new design for cantilevered stators with large hub clearances. *J. Turbomach.* **2013**, *135*, 041022. [CrossRef]
20. Wang, Y.; Chen, W.; Wu, C.; Ren, S. Effects of tip clearance size on the performance and tip leakage vortex in dual-rows counter-rotating compressor. *Proc. Inst. Mech. Eng. Part G J. Aerosp. Eng.* **2015**, *229*, 1953–1965. [CrossRef]
21. Zhu, Y.; Luo, J.; Liu, F. Flow computations of multi-stages by URANS and flux balanced mixing models. *Sci. China Technol. Sci.* **2018**, *61*, 1081–1091. [CrossRef]

Article

Fluidic Thrust, Propulsion, Vector Control of Supersonic Jets by Flow Entrainment and the Coanda Effect

Toshihiko Shakouchi * and Shunsuke Fukushima

Graduate School of Engineering, Mie University, Tsu 514-8507, Japan
* Correspondence: shako@mach.mie-u.ac.jp; Tel.: +81-59-231-9384

Abstract: Thrust, propulsion, vector control of supersonic jets has been applied to jet and rocket engines, ejectors, and other many devices. In general, there are two approaches to this type of control, namely mechanical moving systems and fluidic thrust vector control systems without moving parts, with mechanical moving systems being the most common. However, generally speaking, these systems are very complicated, and more simple methods and devices are desired. In this study, an extremely simple method for the thrust vector control of a supersonic jet by a fluidic Coanda nozzle (FC-nozzle) using the entrainment of the surrounding fluid and Coanda effect is newly proposed. The FC-nozzle consists of a pipe nozzle (Pi-nozzle), spacer, and linearly expanded Coanda nozzle (Co-nozzle) with eight suction pipes (Su-pipes) installed to surround the jet from the Pi-nozzle. The jet from the Pi-nozzle flows straight with the entrainment flow of the surrounding fluid. When some Su-pipes are closed, the pressure between the jet and Co-nozzle wall decreases, and subsequently, the jet deflects to the closed side of the Su-pipe and reattaches to the wall by the Coanda effect. The flow characteristics and deflection characteristics of the supersonic jet from the FC-nozzle are examined by the visualized flow pattern using the Schlieren method and measurements of the velocity distribution. As a result, it is shown that by changing the number of Su-pipes and the locations at which they are closed, the deflection angle and circumferential position of the jet from the Pi-nozzle can be easily controlled.

Keywords: thrust vector control; supersonic jet; nozzle; flow entrainment; Coanda effect; schlieren method

Citation: Shakouchi, T.; Fukushima, S. Fluidic Thrust, Propulsion, Vector Control of Supersonic Jets by Flow Entrainment and the Coanda Effect. *Energies* **2022**, *15*, 8513. https://doi.org/10.3390/en15228513

Academic Editors: Bengt Sunden, Jian Liu, Chaoyang Liu and Yiheng Tong

Received: 13 October 2022
Accepted: 11 November 2022
Published: 14 November 2022

1. Introduction

Supersonic jets are widely used in the industrial field, in applications such as jet and rocket engines, ejectors, gas-atomization, and many other devices. There has been significant research on the flow characteristics of supersonic under-expanded jet flows (SSU-jets). Donaldson and Snedeker [1] and Kojima and Matsuoka [2] showed the flow structure of the SSU-Jet by measuring the pressure and velocity distributions, and Nourl and Whitelaw [3], Zapryaggaezet et al. [4], and Shakouchi et al. [5,6] (2019, 2020) investigated the flow characteristics of the SSU-Jet by visualizing its flow pattern with the Schlieren method and measuring the pressure and velocity distributions to show the applications of this device. Shadow et al. [7] investigated the spreading rate of the SSU-jet, and Gutmark et al. [8] studied the aeroacoustics of turbulent jets and noise reduction by the chevron nozzle.

The vector control of jet thrust through propulsion is an important research subject. Thrust vector control of the SSU-jet is mainly applied to the altitude control of aircrafts by a jet engine or rocket engine. There are two approaches in general, namely mechanical moving systems and fluidic thrust vector control systems without moving parts, with mechanical moving systems being the most common. Recent typical examples of such systems are the mechanical thrust vector control system of the jet engine in V/STOL aircrafts and rocket engines. However, this is a very complicated system, and simpler methods and devices are desired. Some fluidic systems without moving parts have been proposed. Deere [9] summarized the research on a fluidic thrust vectoring system without moving parts conducted at the NASA Langley Research Center and showed that the throat shifting method provides the most efficient thrust control of the fluidic thrust vectoring methods, but larger thrust vector angles can be obtained with the shock vector control method. Additionally, Páscoa et al. [10] reviewed studies on thrust-vectoring in support of a V/STOL non-moving mechanical propulsion system and introduced the idea that, generally speaking, there are three major systems based on mechanical approaches: the two main bulk streams of the Coanda nozzle and the fluidic control concepts. Fluidic vector thrust control systems with co-flow [11,12], counter-flow [13], a synthetic jet actuator [14,15], shock vector control [16], and sonic throat skewing [17] were introduced and explained. Allen and Smith [18] presented a jet vectoring technique with a novel spray method called Coanda-assisted spray manipulation. The primary jet flows through the center of a rounded collar. The control jet is parallel to the primary jet flow and is adjacent to the convex collar according to known Coanda effect principles and entraining and vectoring of the primary jet, resulting in controllable radial-peripheral directional spraying. However, further research on the thrust vector control of supersonic jets is required. Porhashem et al. [19] and Yan, et al. [20] investigated thrust control by fluidic injection and the flow field characteristics of the SSU-jet. Trancossi et al. [10,21–24] analyzed the use of the aerial Coanda high-efficiency orienting nozzle (ACHEON) for aircraft propulsion based on the dynamic equilibrium of two jet streams. Conventional fluidic vector control systems without moving parts also require seconday control jets.

In this paper, a new and simple method for the vector control of supersonic jets is proposed. This method involves a fluidic Coanda nozzle (FC-nozzle) and uses the entrainment of the surrounding fluid and the Coanda effect [6]. The FC-nozzle consists of a pipe nozzle (Pi-nozzle), spacer, and linearly expanded Coanda nozzle (Co-nozzle) with eight suction pipes (Su-pipes) installed to surround the jet from the Pi-nozzle. The jet from the Pi-nozzle flows straight with entrainment of the surrounding fluid. When some suction pipes are closed, the pressure between the jet and Co-nozzle wall decreases, and subsequently, the jet deflects to the closed side of the Su-pipe and attaches to the wall by the Coanda effect. The flow characteristics and deflection characteristics of the supersonic jet from the FC-nozzle were investigated by the visualized flow pattern using the Schlieren method and measurements of the velocity distribution. We show that by changing the number of Su-pipes and the locations at which they close, the deflection angle and circumferential position of the jet can be controlled. Furthermore, when there is no fluid to be entrained by the jet, that is, when the surroundings are in a vacuum state, the same type of control can be performed by blowing small amounts of fluid and air from the Su-pipe. This new fluidic thrust vector control system of the supersonic jet is extremely simple compared to other conventional methods that require control with additional devices, such as secondary jets.

2. Experimental Apparatus and Procedure

2.1. Schematics of the Fluidic Coanda Nozzle

Figure 1 shows a schematic diagram of the fluidic Coanda nozzle (FC-nozzle) with no moving parts. The FC-nozzle is an extremely simple system that consists of a pipe nozzle (Pi-nozzle), spacer, and linearly expanded Coanda nozzle (Co-nozzle) with eight suction pipes (Su-pipes) installed to surround the jet from the Pi-nozzle. The diameter of the Pi-nozzle is d_0 = 4.0 mm, and the inlet diameter, length, and expansion angle of the Co-nozzle are d_{ic} = 5.0 mm, L_c/d_{is} = 2.4, and α = 20°, respectively. On the circumference of the Co-nozzle, eight Su-pipes (n = 8) with diameters of d_{is} = 2.4 mm are installed. The thickness of the spacer is 2.0 mm, and the inlet and outlet diameters are 4.0 and 4.7 mm, respectively. The origin of the coordinates is at the center of the Pi-nozzle exit, and the direction of jet flow is along the x axis. The directions perpendicular to it are the r and r′ axes, as shown in Figure 1.

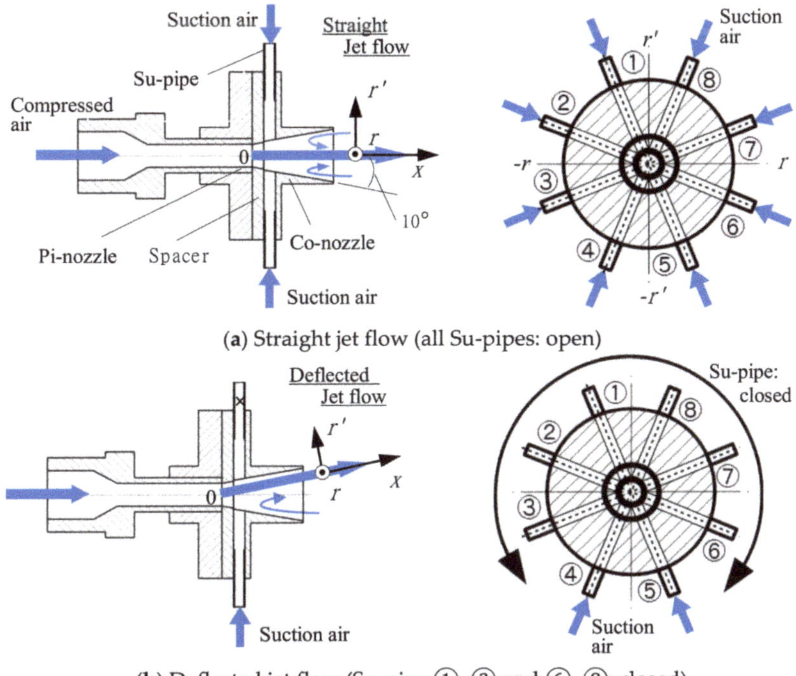

Figure 1. FC-nozzle (Fluidic thrust vector control of the Pi-jet by the FC-nozzle with 8 SU-pipes): When all Su-pipes are opened, the jet flows downstream along the nozzle axis, as shown in (**a**). When some Su-pipes are closed, the pressure between the outer edge of the jet and the wall of the Co-nozzle decreases and then the jet deflects to the closed side of Su-pipe by the Coanda effect as shown in (**b**).

2.2. Experimental Procedure

The compressed air from the air compressor is supplied to the Pi-nozzle after passing through a dryer and flow control valve, and supersonic jets are issued from the Pi-nozzle or FC-nozzle to the ambient air. A static pressure hole with a diameter of 0.8 mm is installed on the pipe wall connected to the upstream side of the Pi-nozzle, and the pressure there is used as the supply pressure, P_0. The supplied pressure changes from P_0 = 0.1 to 0.48 MPa. In the current study, most of the experiments were carried out at a supply pressure of P_0 = 0.38 MPa, which is sufficient to obtain a supersonic jet. The flow pattern was visualized by the Schlieren method (KATO KOKEN, ss150, Japan) and the image was recorded. The total and static pressures in the flow field were measured with Pitot tubes, and the velocity distribution was obtained using their values, taking into account the effect of air compressibility. The outer and inner diameters of the total pressure tube were 1.0 mm and 0.8 mm, respectively, and the diameter and hole diameter of the static pressure tube were 1.0 mm and 0.6 mm, respectively.

3. Propulsion Vectoring Method

When all 8 Su-pipes are opened, the jet entrains the surrounding fluid from the Su-pipes and around the jet, and flows downstream along the nozzle axis, as shown in Figure 1a. On the other hand, when some Su-pipes are closed, the jet cannot entrain surrounding fluid from their sides, and subsequently, the pressure between the outer edge of the jet and the wall of the Co-nozzle decreases, and then, the jet deflects to the closed side of the Su-pipe by the Coanda effect, as shown in Figure 1b. Therefore, arbitrary circumferential vector control of the jet from the Pi-nozzle can be performed by changing the number of Su-pipes and the locations at which they are closed. If these Su-pipes are opened and closed using a solenoid valve and a computer control system, dynamic thrust vector control becomes possible.

There does not necessarily have to be eight Su-pipes. All that is necessary is to block part of the circumference of the jet from the Pi-nozzle.

4. Results and Discussion

First, the flow characteristics of under-expanded supersonic jet from the Pi-nozzle are investigated using the visualized flow pattern and measurements of the velocity distribution.

4.1. Jet Flow from the Pi-Nozzle

4.1.1. Visualized Flow Pattern

The visualized flow pattern of the under-expanded supersonic jet from the pipe nozzle (Pi-nozzle) (Figure 1) produced by the Schlieren method is shown in Figure 2. The supplied pressure is P_0 = 0.38 MPa. The black and white colored areas in the jet are compression and expansion regions, respectively, and white colored x shaped pseudo-shock waves can be seen in the shock cell. The dotted lines in the figure indicate the approximate jet boundary derived from the measurements of the cross sectional velocity distributions at x/d_0 = 2.1, 4.2, 5.4, 8.3. The jet boundary is at the position of r at u/u_c = 0.1. The jet expands at the nozzle exit and the jet width increases almost linearly in the range of x/d_0 > 3.0.

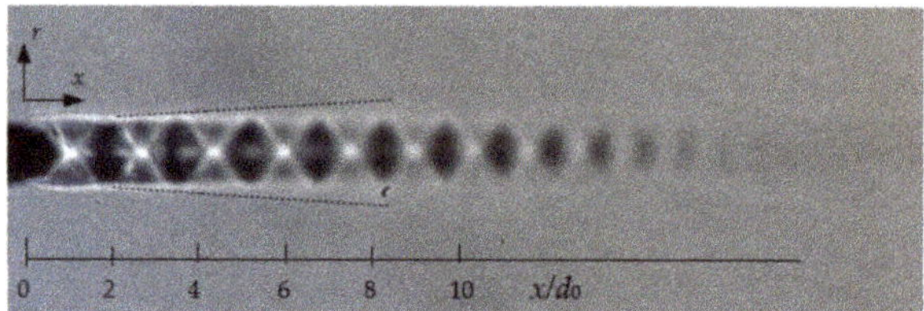

Figure 2. Visualized flow pattern of the under-expanded supersonic jet issued from the Pi-nozzle produced by the Schlieren method, P_0 = 0.380 MPa: The dotted lines indicate the approximate jet boundary with the position of r at u/u_c = 0.1.

4.1.2. Jet Centerline Velocity Distribution

The jet centerline velocity u_c is indicated by a circle in Figure 5 shown later. This was obtained from the measurements of the total and static pressure distributions. The jet repeatedly expands and compresses and reaches the maximum value at x/d_0 = 5.8. In the downstream area of x/d_0 > 12.5, u_c decay occurs slowly.

4.2. Jet Flow from the FC-Nozzle

Next, the flow characteristics and vector control of the jet issued from the FC-nozzle with the Co-nozzle were investigated using the visualized flow pattern and measurements of the velocity distribution.

4.2.1. Visualized Flow Pattern

Figure 3a shows the visualized flow pattern of the FC-jet issued from the FC-nozzle when all the Su-pipes are opened, 0-close. The supplied pressure is P_0 = 0.38 MPa. The jet flows straight to the downstream entraining the surrounding fluid from the Su-pipes and around the jet (Figure 1a). The shock waves can be seen in the jet, and the white and black colored areas from the nozzle exit represent the expansion and compression shock waves, respectively. However, the FC-jet with 0-close decays faster than the Pi-jet shown in Figure 2, because the entrainment of surrounding fluid is suppressed by the Co-nozzle.

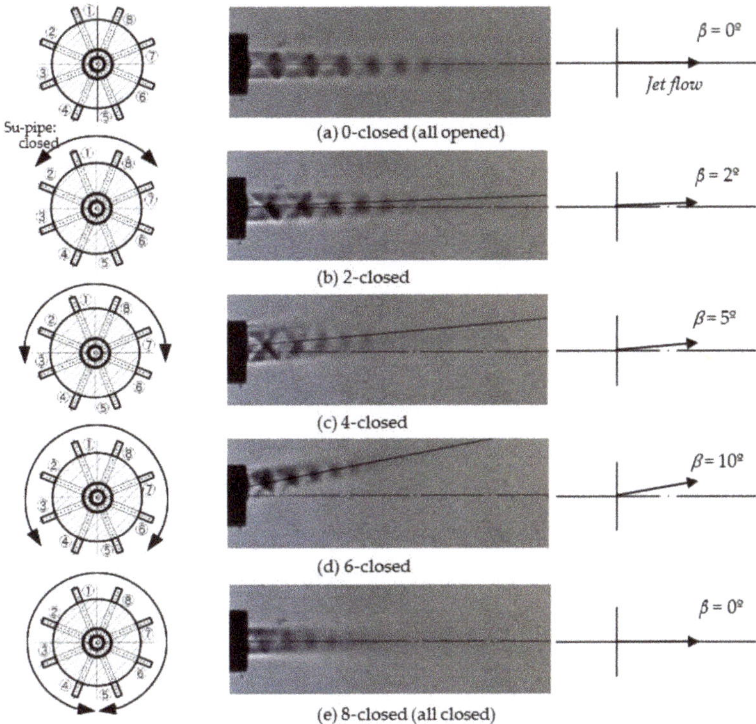

Figure 3. Visualized flow pattern of the jet issued from the FC-nozzle by Schlieren method, P_0 = 0.38 MPa (Effect of suction flow): The deflection angle of the jet was measured using visualized flow pattern.

Figure 3d shows the case in which Su-pipes ①–③ and ⑥–⑧ are closed (6-closed). The entrainment from those Su-pipes is prevented, and the pressure between the outer edge of the jet from the Pi-nozzle and the wall of the Co-nozzle decreases. Consequently, the jet deflects and reattaches to the wall of the Co-nozzle by the Coanda effect. The deflection angle is $\beta = 10°$, which is the same as the opening angle of the Co-nozzle.

Figure 3b,c show the 2- and 4-closed cases, respectively. Figure 3e shows the 8-closed case, where the jet flows straight to the downstream area; however, since the jet cannot entrain the surrounding fluid from the Su-pipe and the flow resistance of the nozzle increases, the jet velocity is slower than those of the others.

4.2.2. Deflection Angle, β

Figure 4 shows the deflection angle, β, and the number of closed Su-pipes, n_c. β increases as n_c increases and has a maximum value of $\beta = 10°$ at $n_c = 6$. In the figure, the results for the greater supply pressure, P_0 = 0.46 MPa, are also shown for reference, and both values are approximately the same. By changing the number of Su-pipes and the locations at which the Su-pipes close, the deflection angle and circumferential position of the jet can be controlled.

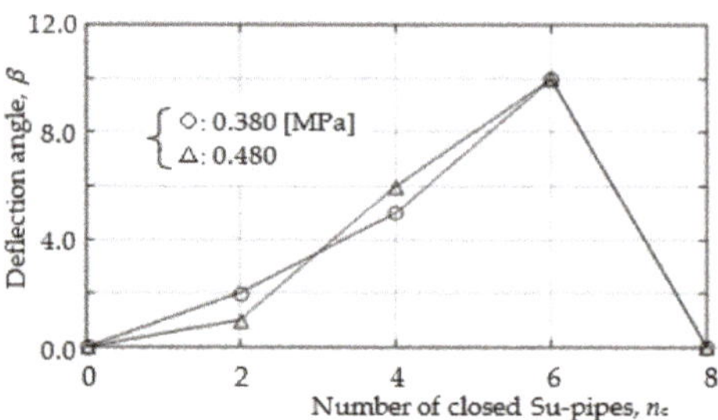

Figure 4. Deflection angle, β (P_0 = 0.38, 0.48 MPa): The deflection angle has a maximum value of β = 10° at n_c = 6.

4.2.3. Practical Operation

In practical applications, it is conceivable to use a computer-controlled solenoid valve to change the number of Su-pipes and the locations at which they close.

4.3. Velocity Distribution

In order to clarify the flow characteristics of the straight and deflected jets, the jet center line and cross-sectional velocity distributions were measured.

4.3.1. Jet Centerline Velocity Distribution

Figure 5 shows the centerline velocity distribution of the jets u_c at P_0 = 0.38 MPa. The velocity was obtained by measurements of the total and static pressure distributions with consideration of the compressibility of the fluid, air. In Figure 5, the position of the nozzle exit of the FC-nozzle is x/d_0 = 3.5 (the distance in the x-direction from the exit of the FC-nozzle: s/d_0 = 0.0). The Pi-jet flows downstream with a large fluctuation of u_c, corresponding to the expansion and compression shock waves of the flow, and u_c decreases monotonously in the region of $x/d_0 > 11$. In the 0-closed case, u_c is smaller than in the Pi-jet due to the large flow resistance of the Co-nozzle. Moreover, as the number of closed Su-pipes increases, the entrainment of the surrounding fluid from the Su-pipe decreases. The u_c of the 8-closed condition decreases monotonously, and it approximately equals that of the 6-closed case in the region of $x/d_0 > 8$.

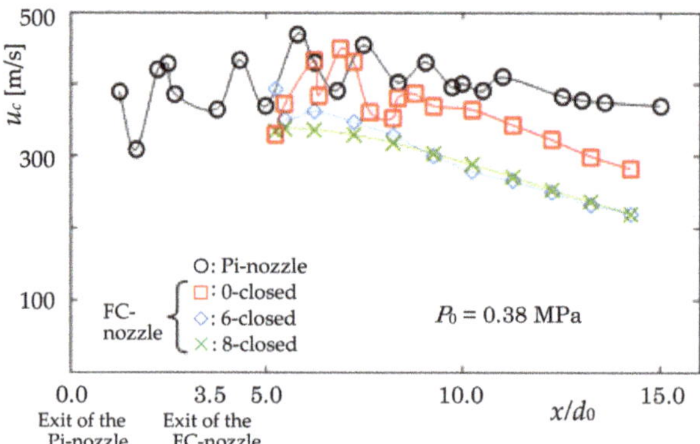

Figure 5. Jet centerline velocity distribution, u_c (0-, 6-, 8-closed jets and Pi-jet, P_0 = 0.380 MPa): The u_c of the 0-closed nozzle is smaller than that of the Pi-jet owing to the large flow resistance of the Co-nozzle.

4.3.2. Cross-Sectional Pressure and Velocity Distributions

1. Total Pressure Distribution

Figure 6a shows the total pressure distributions, P_t/P_0, in the cross-section at $x/d_0 = 4.2$ ($s/d_0 = 0.7$) for the 0-closed condition as an example. The jet of the 0-closed case flow straight to the downstream area, and the profile is axisymmetric and dented at the center. In comparison, the 6-closed case has a small maximum total pressure because of the large flow resistance of the Co-nozzle and the asymmetric profile.

Figure 6b,c are the results at $x/d_0 = 5.4$, 8.3, respectively. The profile at $x/d_0 = 8.3$ for the 0-closed case is axisymmetric, and the maximum value is approximately twice that of the 6-closed case.

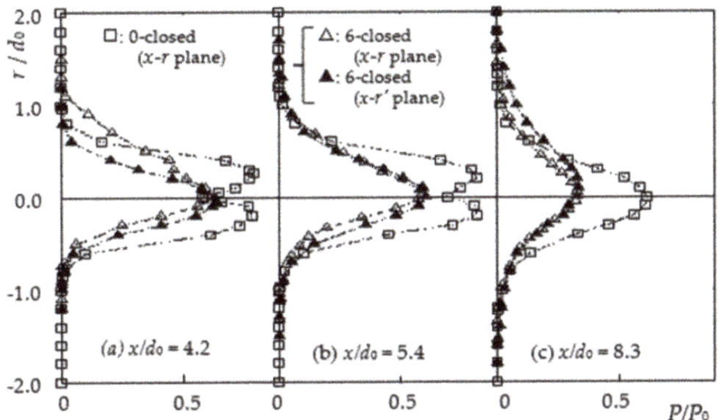

Figure 6. Total pressure distribution, P/P_0, in the cross section (0−, 6−closed jets, P_0 = 0.380 MPa).

2. Static Pressure Distribution

Figure 7a shows the static pressure distribution, P_s/P_0, in the cross section at $x/d_0 = 5.4$ ($s/d_0 = 1.9$) for the 0-closed case. The profile is dented at the center by the expansion shock wave [2].

Figure 7b shows the results for the 6-closed case and Figure 7c is a three-dimensional drawing of Figure 7b. The profile is asymmetric, because the jet deflects and reattaches to the wall of the Co-nozzle, and consequently, the shape of the jet deforms. The jet reattaches to the wall of the Co-nozzle at the $+r'$-axis side and entrains the surrounding fluid from the $-r'$-axis side because of negative pressure.

3. Sectional Velocity Distribution

Figure 8 shows the velocity distribution obtained with the total and static pressure measurements under the consideration of the compressibility of the fluid in the cross-section at $x/d_0 = 5.4$. The velocity profile of the 6–closed case is smaller than that of the 0-closed case because of a larger flow resistance of the Co-nozzle.

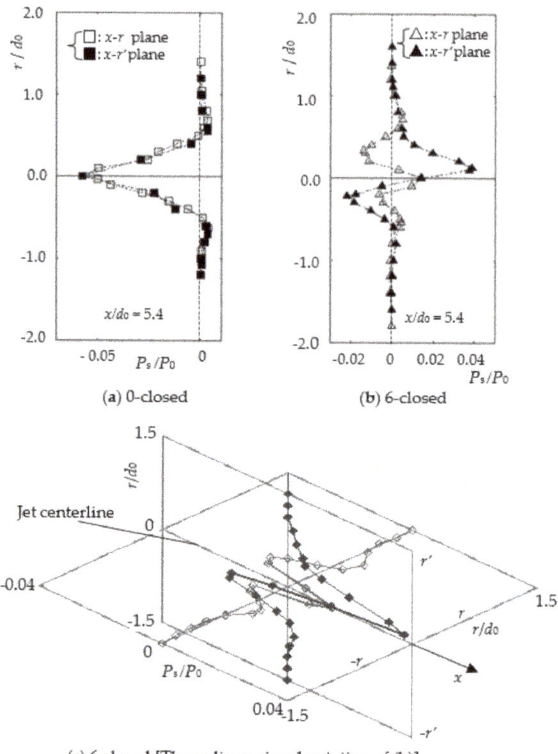

Figure 7. Static pressure distribution, P_s / P_0, in the cross section at x/d_0 = 5.4 (P_0 = 0.38 MPa): The jet reattaches to the wall of the Co-nozzle at the $+r'$−axis side and entrains the surrounding fluid from the $-r'$-axis side of the negative pressure.

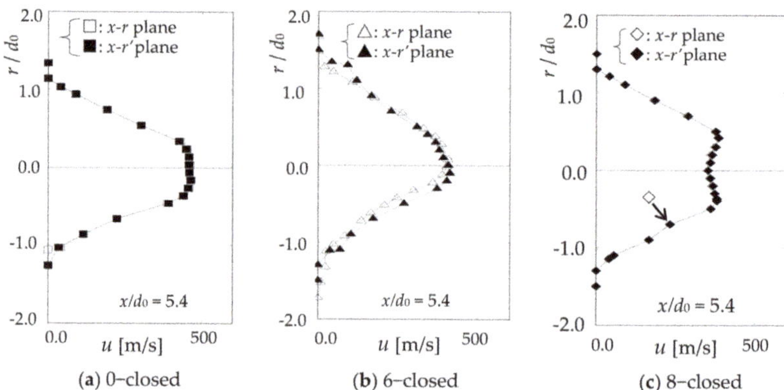

Figure 8. Cross sectional velocity distribution at x/d_0 = 5.4 (P_0 = 0.38 MPa): The velocity profile of the 6−closed nozzle is asymmetric because the jet deflects and reattaches to the side wall of the Co-nozzle.

Figure 9 shows the velocity distribution in the cross section at x/d_0 = 4.2, 5.4, and 8.3 for the 0-, 6-, and 8-closed cases. In Figure 9a, the results for the Pi-jet are also shown for reference. The velocity was obtained by taking the static pressure, P_s, as the ambient pressure, i.e., the atmospheric pressure, P_a. The velocity distribution of the 0-closed case at

x/d_0 = 5.4 and the results shown in Figure 8a are approximately the same, because the static pressure is much smaller than the total pressure, as shown in Figures 6 and 7. In this case, it seems that the velocity may be obtained by using P_s as P_a.

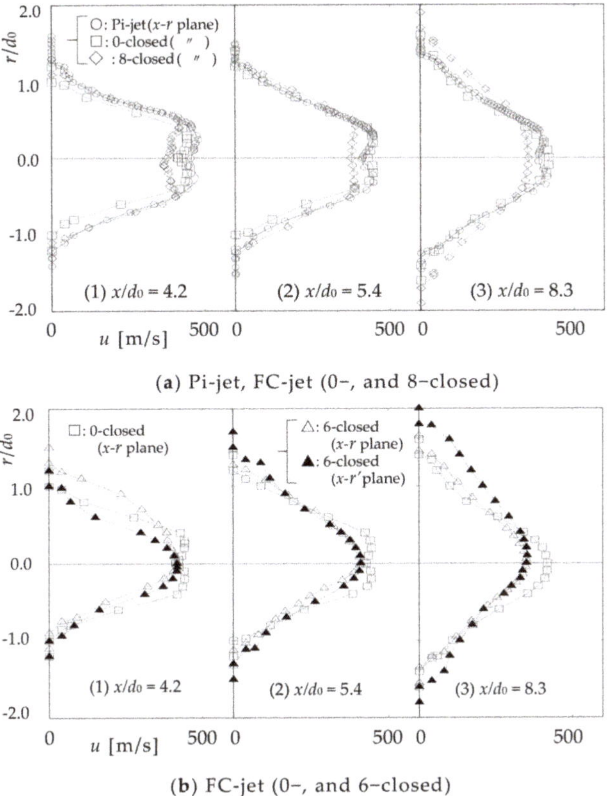

Figure 9. Cross sectional velocity distribution at x/d_0 = 4.2, 5.4, 8.3 (P_0 = 0.38 MPa, P_s = P_a): The maximum velocity of the 0−closed case is approximately equal to that of the Pi-jet and is larger than that of the 8−closed case.

The maximum velocity of the 0-closed case is approximately equal to that of the Pi-jet and larger than that of the 8-closed case, because the flow resistance of the Co-nozzle for the 8-closed case is greater than those of the others.

The flow characteristics of the straight and deflected flows were made clear by the velocity measurements.

4.3.3. Thrust, Approximate Estimation

The deflected jet has an asymmetric velocity profile, as shown in Figure 9b. The thrust, F, of the jet was calculated by the following equation under the assumption that the cross-section of the jet is approximately circular. At this time, the mean values of the x-r and x-r' planes were taken for the asymmetric velocity distribution. The density, ρ, was estimated approximately using the equation of the state of gas.

$$F = \int_0^\infty \rho u^2 2\pi r dr + (P_e - P_a) A_e \quad (1)$$

Figure 10 shows the thrust, F, of the FC-jet calculated at x/d_0 = 5.4 (s/d_0 = 1.9) for P_0 = 0.38 MPa. The F of the 6-closed case is smaller than that of the 0-closed case because the entrainment from the surroundings of the 6-closed case is restricted. On the other hand, the F of the 8-closed case is larger than that of the 0-closed case, and this seems to be due to the jet approaching proper expansion. The differences in the F value when the static pressure is atmospheric pressure, and when the measurements were used, the value was ± 6 % for the 0- and 8-closed cases and ±29% for the 6-closed case, respectively. The large difference for the 6-closed case is probably due to the deformation and asymmetry of the jet's shape. We showed that by changing the number of the Su-pipes and the locations at which they close, the deflection angle and circumferential position of the jet can be controlled, and consequently, the thrust of the supersonic jet can be controlled by a new simple fluidic vector control system.

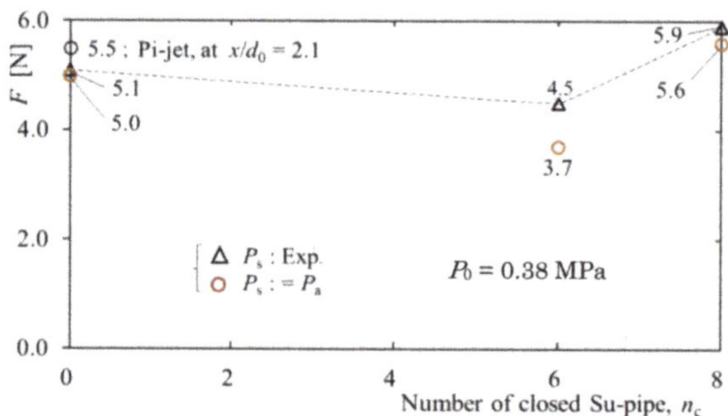

Figure 10. Thrust, F, of the supersonic jet from the FC-nozzle: The thrust of the 8-closed case is greater than that of the 0-closed case, and this seems to be due to the jet approaching proper expansion.

4.3.4. Thrust Vectoring by Sub Jets

When there is no fluid to be entrained by the jet, that is, when the surrounding area is in a vacuum state, the same control can be performed by blowing a small amount of fluid from the Su-pipe.

Using the same nozzle, the FC-nozzle, to blow sub-jets from the two Su-pipes (2-blow) at a small portion of the total flow rate of Q_c = 10 [L/min], the jet supplied with P_0 = 0.38 MPa is deflected, similar to the result shown in Figure 3d, at an angle of β = 10° to the side that does not blow out, because the pressure on the blowing side increases. Q_c = 10 [L/min] is about 1.8% of the main jet flow estimated under the standard state.

Figure 11 shows the visualized flow pattern, and the number of Su-pipes blowing is n_b = 2 from ④ and ⑤ in Figure 3 with all remaining Su-pipes are closed.

1. Effect of the Blowing Flow Rate Q_c on the Deflection Angle β Figure 12 shows the relationship between β and Q_c when air is blown from Su-pipes 2, 4, and 6. β increases as Q_c increases, and β also increases as the number of blowing nozzles, n_b, decreases, because by increasing n_b, the blowing flow rate per Su-pipe decreases. The deflection angle has a maximum value of β = 10° at n_b = 2 and Q_c = 8 [L/min].

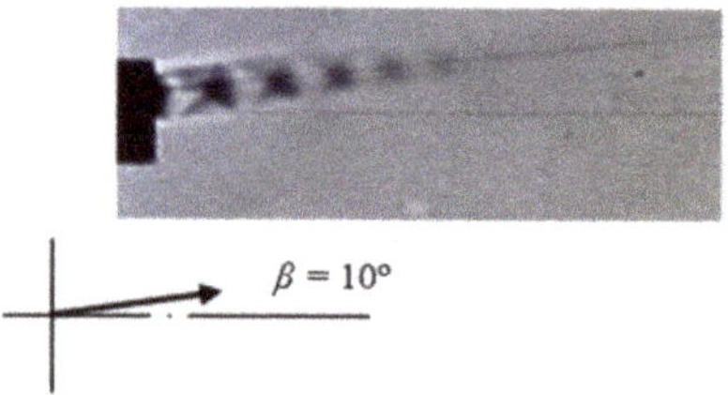

Figure 11. Deflection angle, β, of the 2-blow condition (P_0 = 0.38 MPa, Q_c = 10 [L/min]).

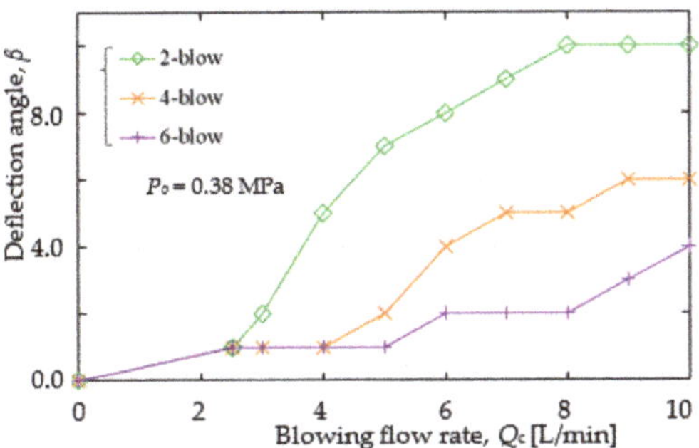

Figure 12. Deflection angle, β and blowing flow rate, Q_c (P_0 = 0.38 MPa): The deflection angle increases as Q_c increases, and β also increases as the number of blowing nozzle n_b decreases.

2. Effect of the Number of Blowing Nozzles, n_b, on the Deflection Angle β

Figure 13 indicates the relation between β and n_b. The deflection angle has a maximum value of $\beta = 10°$ at $n_b = 2$, and in the 4−, 6−, and 8−blow cases, the deflected angles are 6°, 4°, and 0°, respectively.

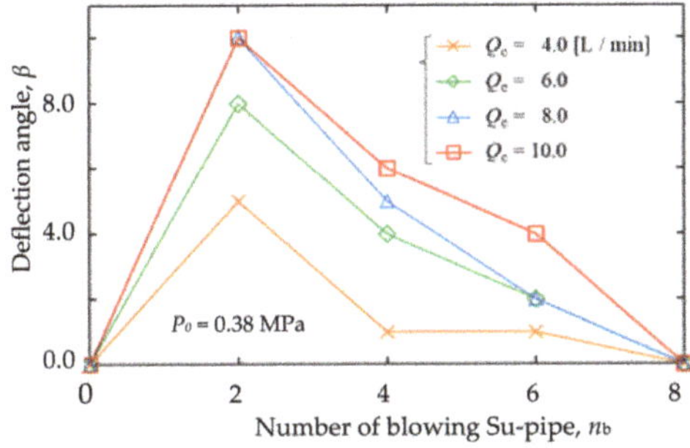

Figure 13. Deflection angle, β, and the number of Su-pipes blowing, n_b (P_0 = 0.38 MPa, Q_c = 10 [L/min]): The deflection angle has a maximum value of $\beta = 10°$ at $n = 2$.

It is considered that the deflection angle β can be increased by increasing the opening angle α of the FC-nozzle. However, investigation of the characteristics of the shape of the FC-nozzle, such as the optimal length and α, is a subject for further study.

5. Conclusions

In this paper, thrust vector control of the supersonic under-expanded jet by a non-moving system was examined, and a new and simple vector control method for supersonic jets was proposed. The method uses a fluidic Coanda nozzle (FC-nozzle) entrain the surrounding fluid and the Coanda effect. The FC-nozzle consists of a pipe nozzle (Pi-nozzle), spacer, and linearly expanded Coanda nozzle (Co-nozzle) with eight suction pipes (Su-pipes). The jet from the Pi-nozzle flows straight with entrainment of the surrounding fluid from the Su-pipes and around the jet. When some Su-pipes are closed, the pressure between the jet and Co-nozzle wall decreases, and subsequently, the jet deflects to the closed side of the Su-pipe and attaches to the wall of the Co-nozzle by the Coanda effect. The flow characteristics and deflection characteristics of the supersonic jet issued from the FC-nozzle were assessed by the visualized flow pattern using the Schlieren method and measurements of the velocity distribution.

(1) By changing the number of the Su-pipes and the locations at which the Su-pipes close, the deflection angle and circumferential position of the jet can be controlled. That is,

vector control of the thrust of the supersonic jet can be performed with this new and simple method and apparatus.

(2) This new fluidic thrust vector control method for supersonic jets is extremely simple compared to other conventional methods that require control with additional devices, such as secondary jets.

(3) In practical applications, it is conceivable that a computer-controlled solenoid valve could be used to change the number of Su-pipes and the locations at which they close.

(4) When there is no fluid to be entrained by the jet, that is, when the surroundings are in a vacuum state, the same type of control can be gained by blowing a small amountof fluid from the Su-pipe.

The effects of the expansion angle and length of the Co-nozzle on jet deflection are issues for further study.

Author Contributions: Conceptualization, T.S.; methodology, T.S.; software, T.S. and S.F.; validation, T.S. and S.F.; formal analysis, T.S.; investigation, T.S.; resources, T.S.; data curation, T.S. and S.F.; writing—original draft preparation, T.S.; writing—review and editing, T.S. and S.F.; visualization, T.S.; supervision, T.S.; project administration, T.S. All authors have read and agreed to the published version of the manuscript.

Funding: This research received no external funding.

Informed Consent Statement: Not applicable.

Data Availability Statement: Not applicable.

Acknowledgments: The authors would like to thank K. Tsujimoto, T. Ando and M. Takahashi for their valuable discussions and Messrs. T. Tanoue and K. Hinaga for their cooperation with the experiments.

Conflicts of Interest: The authors declare no conflict of interest.

Nomenclature

A_e	exit area of the Pi-nozzle
d_{ic}	diameter of the Su-pipe
d_{is}	inlet diameter of the Co-nozzle
d_0	diameter of the Pi-nozzle
F	thrust
L_c	length of the Co-nozzle
n	number of Su-pipes
n_b	number of blowing Su-pipes
n_c	number of closing Su-pipes
P_a	ambient pressure (atmospheric pressure)
P_e	exit pressure of the Pi-nozzle
P_s	static pressure
P_t	total pressure
P_0	supply pressure
Q_c	total blowing flow rate
r, r'	perpendicular directions of the x-axis, respectively (Figure 1)
s	distance in the x-direction from the exit of the FC-nozzle
u	velocity in the x direction
u_c	jet centerline velocity
x	direction of jet flow (Figure 1)
α	expansion angle of Co-nozzle
β	jet deflection angle
ρ	density of fluid (air)

Abbreviations

Co-nozzle	Coanda nozzle
FC-nozzle	fluidic Coanda nozzle
Pi-nozzle	pipe nozzle
SSU-jet	supersonic under-expanded jet flow
Su-pipe	suction pipe

References

1. Donaldson, S.D.; Snedeker, R.S. Abst: Under-Expanded Jet. 1971. Available online: http://www.kbwiki.ercoftac.org/w/index.pkp/Abstr:Under-expanded_jet (accessed on 1 March 2019).
2. Kojima, T.; Matsuoka, Y. Study of an under-expanded sonic jet (1st Report, On the mechanism of pseudo-shock waves and pressure and velocity in jets). *Trans. Jpn. Soc. Mech. Eng. JSME* **1987**, *259*, 190–191. (In Japanese)
3. Nourl, J.M.; Whitelaw, J.H. Flow characteristics of an under-expanded jet and its application to the study of droplet breakup. *Exp. Fluids* **1996**, *21*, 243–247. [CrossRef]
4. Zapryagaez, V.I.; Kiselev, N.P.; Pivovarov, A.A. Gas dynamic structure of an axisymmetric supersonic under-expanded jet. *Fluid Dyn.* **2015**, *50*, 95–107.
5. Shakouchi, T.; Tanoue, T.; Tsujimoto, K.; Ando, T. Flow characteristics of sub- and supersonic jets issued from orifice and notched orifice nozzles, and application to ejector. *J. Fluid Sci. Technol. JSME* **2019**, *14*, 1–12. [CrossRef]
6. Shakouchi, T. Passive flow control of jets. *J. Mech. Eng. Rev. JSME* **2020**, *17*, 1–18. [CrossRef]
7. Shadow, K.C.; Gutmark, E.; Wilson, W.J. Compressible spreading rates of supersonic coaxial jets. *Exp. Fluids* **1990**, *10*, 161–167. [CrossRef]
8. Gutmark, E.J.; Callender, B.W.; Martens, S. Aeroacoustics of Turbulent Jets: Flow Structure, Noise Sources, and Control. *JSME Int. J. Ser. Fluid Therm. Eng.* **2006**, *49*, 1078–1085. [CrossRef]
9. Deere, K.A. Summary of fluidic thrust vectoring research conducted at NASA Langley Research Center. In Proceedings of the 21st AIAA Applied Aerodynamics Conference, Orlando, FL, USA, 1 January 2003; AIAA-2003-3800. pp. 1–18.
10. Pascoa, L.; Dumas, A.; Trancossi, M.; Stewart, P.; Vucinic, D. A review of thrust vectoring in support of V/STOL non-moving mechanical propulsion system. *Cent. Eur. J. Phys.* **2013**, *3*, 374–388. [CrossRef]
11. Heo, J.Y.; Too, K.H.; Lee, Y.; Sung, H.G.; Cho, S.H.; Jeon, Y.J. Fluidic thrust vector control of supersonic jet using co-flow injection. In Proceedings of the 45th AIAA/ASME/SAE/ASEE Joint Propulsion Conference Exhibit, Denver, CO, USA, 2–5 August 2009; AIAA 2009-517. pp. 1–2.
12. Dumitrache, A.; Frunzulica, F.; Preotu, O.; Ionescu, T. Thrust and jet directional control using the Coanda effect. *IN-CAS Bull.* **2018**, *10*, 199–205. [CrossRef]
13. Wu, K.; Kim, H.D.; Jin, Y. Fluidic thrust vector control based on counter-flow concept. *Proc. Inst. Mech. Eng. Part G J. Aerosp. Eng.* **2018**, *233*, 1412–1422. [CrossRef]
14. Trancossi, M.; Dumas, A. *Coanda Synthetic Jet Deflection Apparatus and Control*; No. 2011-01-2590; SAE International: Warrendale, PA, USA, 2011. [CrossRef]
15. Liu, J.F.; Luo, Z.B.; Deng, X.; Zhao, Z.J.; Li, S.Q.; Liu, Q.; Zhu, Y.X. Dual Synthetic jets actuator and its applications—Part II: Novel fluidic thrust-vectoring method based on dual synthetic jets actuator. *Actuators* **2022**, *11*, 209. [CrossRef]
16. Resta, E.; Malsilio, R.; Ferlauto, M. Thrust vectoring of a fixed axisymmetric supersonic nozzle using the shock-vector control method. *Fluids* **2021**, *6*, 441. [CrossRef]
17. Miller, R.; Yagle, P.J.; Hamstra, J.W. Fluidic throat skewing for thrust vectoring in fixed-geometry nozzles. In Proceedings of the 37th Aerospace Sciences Meeting and Exhibit, Reno, NV, USA, 11–14 January 1999; AIAA-99-0365. [CrossRef]
18. Allen, D.; Smith, B.L. Axisymmetric Coanda-assisted vectoring. *Exp. Fluids* **2009**, *46*, 55–64. [CrossRef]
19. Pourhashem, H.; Kumar, S.; Kalkhoran, I.M. Flow field characteristics of a supersonic jet influenced by downstream micro jet fluidic injection. *Aerosp. Sci. Technol. AESCTE* **2019**, *93*, 1–15. [CrossRef]
20. Yan, D.; Wei, Z.; Xie, K.; Wang, N. Simulation of thrust control by fluidic injection and pintle in a solid rocket motor. *Aerosp. Sci. Technol. AESCTE* **2020**, *99*, 1–10. [CrossRef]
21. Trancossi, M. An overview of Scientific and Technical Literature on Coanda Effect Applied to Nozzles. *SAE Tech. Pap.* **2011**, *2011*, 2591. [CrossRef]
22. Trancossi, M.; Dumas, A. *ACHEON: Aerial Coanda High Efficiency Orienting-Jet Nozzle*; No. 2011-01-2737; SAE International: Warrendale, PA, USA, 2011. [CrossRef]
23. Trancossi, M.; Madonia, M.; Dumas, A.; Angeli, D.; Bingham, C.; Sumanta Das, S.; Grimaccia, F.; Marques, J.P.; Porreca, E.; Smith, T.; et al. A new aircraft architecture based on the ACHEON Coanda nozzle: Flight model and energy evaluation. *Eur. Transp. Res. Rev.* **2016**, *8*, 1–21. [CrossRef]
24. Trancossi, M.; Stewart, J.; Maharshi, S.; Angeli, D. Mathematical model of a constructal Coanda effect nozzle. *J. Appl. Fluid Mech.* **2016**, *9*, 2813–2822. [CrossRef]

Perspective

Progress of Coupled Heat Transfer Mechanisms of Regenerative Cooling System in a Scramjet

Ni He [1], Chaoyang Liu [1,*], Yu Pan [1] and Jian Liu [2,*]

1 College of Aerospace Science and Engineering, National University of Defense Technology, Changsha 410073, China
2 Research Institute of Aerospace Technology, Central South University, Changsha 410012, China
* Correspondence: liuchaoyang08@nudt.edu.cn (C.L.); jian.liu@csu.edu.cn (J.L.)

Abstract: The feasibility of regenerative cooling technology in scramjet engines has been verified, while the heat transfer behavior involved in the process needs further study. This paper expounds on the necessity of coupled heat-transfer analysis and summarizes its research progress. The results show that the effect of pyrolysis on heat transfer in the cooling channel depends on the heat flux and coking rate, and the coupling relationship between combustion and heat transfer is closely related to the fuel flow rate. Therefore, we confirm that regulating the cooling channel layout according to the real heat-flux distribution, suppressing coking, and accurately controlling the fuel flow rate can contribute to accomplishing the optimal collaborative design of cooling performance and combustion performance. Finally, a conjugate thermal analysis model can be used to evaluate the performance of various thermal protection systems.

Keywords: scramjet; regenerative cooling; coupled heat transfer; pyrolysis

Citation: He, N.; Liu, C.; Pan, Y.; Liu, J. Progress of Coupled Heat Transfer Mechanisms of Regenerative Cooling System in a Scramjet. *Energies* **2023**, *16*, 1025. https://doi.org/10.3390/en16031025

Academic Editor: Dmitry Eskin

Received: 5 December 2022
Revised: 11 January 2023
Accepted: 14 January 2023
Published: 17 January 2023

1. Introduction

Hypersonic vehicles are a major development direction for the aeronautic and astronautic industries in the 21st century [1]. The constantly improved flight speed of hypersonic vehicles must be accompanied by a sharp increase in aerodynamic heating [2]. Scramjet is considered a promising power device for hypersonic vehicles because of its simple structure and high specific impulse [3]. Fast and efficient combustion is an inevitable prerequisite for the operation of scramjet engines [4]. However, the aerodynamic heating of the approaching airstream and the combustion heat release of fuel will produce enormous thermal loads, making scramjet engines endure severe thermal environment challenges during flight [5]. Therefore, scramjet engines must rely on effective thermal protection systems to ensure their operation stability. In particular, with the continuous increase in the flight Mach number, thermal protection technology is highly coupled with the working processes of the engine gas side and has become a technological bottleneck point limiting the improvements in engine performance.

Regenerative cooling is widely used in liquid rocket engines as a thermal protection technology, and its feasibility in scramjet engines has also been verified. For instance, the regenerative cooling scramjet engine X-51A successfully flew for 140 s [6]. The working process is that the pressurized fuel enters the cooling channel to remove the wall's heat and then injects it into the combustion chamber to complete the combustion [7]. This method can not only cool the engine but also improves the performance of the scramjet [8]. The heat transfer modes involved in regenerative cooling mainly include three modules: convection and radiation heat transfer from the gas to combustion chamber inwall, heat conduction of the combustion chamber structure, and convection heat transfer from the hot wall to the coolant.

This paper summarizes the main characteristics of the regenerative cooling process of a scramjet engine, expounds on the necessity of coupled heat-transfer analysis, and introduces the research progress of the coupled heat-transfer technology. Furthermore, the main problems in the study of coupled heat transfer are discussed to provide a reference for the optimization design and practical application of the thermal protection system of the scramjet engine.

2. Necessity of Coupled Heat Transfer Analysis in Regenerative Cooling

The principal characteristics of the regenerative cooling process of a scramjet engine are as follows:

(1) Fuel in the cooling channel operates in supercritical conditions, mainly to avoid heat transfer deterioration due to phase change, which may lead to the failure of the thermal protection system;
(2) After exceeding the pyrolysis temperature, the endothermic hydrocarbon fuel will produce a pyrolysis reaction to form a mixture of small molecule gases, providing an additional chemical heat-sink;
(3) Fuel composition and physical properties change dramatically during cooling due to the combined action of endothermic and cracking reactions;
(4) Coking may occur during the endothermic pyrolysis of fuel, and the formation of coking deposits will lead to heat transfer deterioration and even channel blockage;
(5) The change in fuel composition will also affect its combustion performance; in addition, the thermal environment of the combustor wall is determined by the combustion performance and cooling efficiency.

Aiming at the thermal environment and the above characteristics of a scramjet engine, scholars have carried out extensive research on aerodynamic heating, supersonic combustion, and supercritical flow heat transfer. However, few studies that have jointly considered these aspects. In terms of the heat transfer characteristics of supercritical fluid in a cooling channel, most of the existing studies have used fixed heat flux boundary conditions. However, the research by Zhao et al. [9] showed that the effects of constant and variable heat flux on the pyrolysis process were different; they considered it necessary to study the coupling mechanism of endothermic hydrocarbon fuel pyrolysis and heat transfer under non-uniform heat flux. In the study of Han et al. [10], the influence of the coupling effect on temperature results was more than 100 K, strongly supporting Zhao and colleagues' conclusion.

In summary, the coupled heat transfer mechanism in the combustor is crucial for scramjet engines' optimal collaborative design.

3. Research Progress of Regenerative Cooling Coupled Heat Transfer in Scramjet

According to the above analysis, the following discussion will include: the pyrolysis heat transfer coupling in the cooling channel, the combustion and heat transfer coupling in the combustion chamber, and the conjugate thermal analysis considering aerodynamic heating in the regenerative cooling of scramjet engines.

3.1. Pyrolysis Heat Transfer Coupling in the Cooling Channel

Figure 1 [11] shows that the fuel in the cooling channel will undergo a chemical reaction after the temperature exceeds the critical temperature, which will further affect the flow and heat transfer process. For this coupling relationship and interaction mechanism, Feng et al. [12,13] found that a chemical reaction boundary exists during the flow development in a tube beside the velocity and temperature boundary layer; in addition, they also established a one-dimensional framework to amplify the coupling relationship between flow and pyrolysis reaction and defined the characteristic time to quantitatively describe its coupling scale.

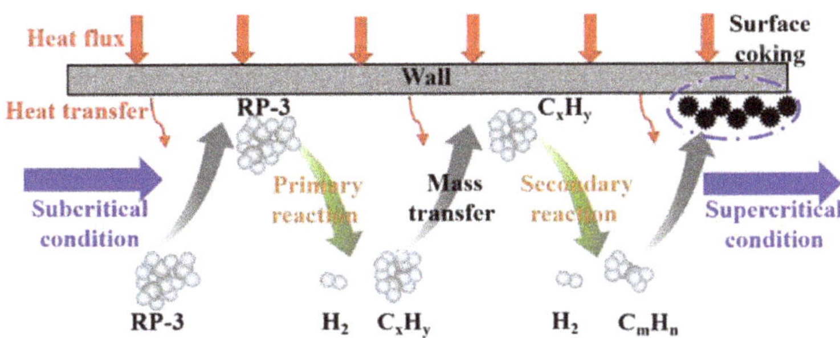

Figure 1. The schematic of heat and mass transfer in the cooling channel [11].

From the perspective of flow characteristics, Jiang et al. [14] numerically analyzed the effect of supercritical n-decane pyrolysis on heat transfer performances. The results showed that the small molecular substances produced by pyrolysis reduced the overall density, the diffusion capacity of fuel components was enhanced, and the inhomogeneity of flow distribution was reduced. Lei et al. [15] analyzed the evolution process of coupled heat transfer in pyrolysis from the perspectives of boundary layer development, turbulent shear stress, and thermophysical properties. The results showed that the heat transfer coefficient was positively correlated with the mass flow rate, negatively correlated with the equivalent diameter, and nonlinearly correlated with the heat flux density. Meanwhile, the effect of heat flux was related to the development of the thermal boundary layer and the change of thermophysical properties caused by pyrolysis. Tian et al. [16] considered that the pyrolysis reaction of RP-3 fuels could increase the velocity at the outlet of the channel by three times. Moreover, Li et al. [17] quantitatively calculated the effect of the cooling channel aspect ratio on heat transfer capacity. The results showed that the increase in the aspect ratio improved the heat transfer performance of the main flow region. The heat transfer coefficient at the aspect ratio of 5 was five times higher than that at the aspect ratio of 3, and the wall temperature at the corner was increased by 4K.

Pyrolysis usually occurs first at high-temperature regions near the wall. Li et al. [18] studied to what purpose this occurred and found that the effect of pyrolysis on heat transfer was nonlinearly related to the variety of heat flux, which was related to the secondary products produced by pyrolysis and the thermal boundary layer effect. The secondary product influenced the fuel composition, causing an increase in viscosity, affecting the thermal diffusivity, and thus reducing the Nusselt number. Meanwhile, the pyrolysis near the wall caused dramatic changes in the radial distribution of the fluid, changed the thermal properties in the boundary layer, and led to the deterioration of heat transfer. Tian et al. [11] also obtained the same conclusion and elaborated from the perspective of hydrocarbon fuel chemical heat sink, flow transition, and ultra-high wall temperature. They reported that the pyrolysis reaction provided a chemical heat sink and accelerated the flow rate, thereby reducing the fuel heating rate and wall temperature. At the same time, the mass fraction of small molecule products increased along with the pyrolysis reaction, and the risk of surface coking and carbon deposition also increased. In addition, Gong et al. [19] numerically studied the effect of the secondary reaction on the heat transfer in reacting flow and reported that the heat transfer efficiency reaches a maximum value with increased hydrocarbon fuel conversion.

Jiao et al. [20,21] experimentally studied the pyrolysis heat-transfer coupling characteristics of RP-3 flowing in a vertical upward tube at supercritical pressures in the context of the buoyancy effect. The results showed that under the condition of high heat flux, the buoyancy effect will lead to heat transfer deterioration which cannot be restored by increased pressure. Pyrolysis promotes the heat transfer performance of fuel by doubling the average thermal conductivity in the pyrolysis zone. Next, they focused on microscopic heat-transfer behavior in the heat transfer deterioration and the cracked region by numerical simulation. They considered that pyrolysis affected heat transfer by two mechanisms: furnish heat absorption or deliver capacity through the endothermic (exothermic) chemical reactions and modify fuel composition and physical properties. The two factors interact and work together to produce different results in different regions. In addition, Li et al. [22] studied the coupling effect of pyrolysis and flow heat transfer from the perspectives of wall temperature distribution, thermal acceleration, n-decane mass fraction, streamline characteristics, and vortex structure. The results showed that pyrolysis reduced the shear stress and turbulent kinetic energy, reducing the velocity difference between the viscous sublayer and the buffer layer, and increasing the near-wall temperature. From another perspective, the high cracking in the near-wall region significantly reduces the density of fuel components, increases the velocity in the boundary layer, and further reduces the wall temperature. This means that thermal acceleration is beneficial to enhance heat transfer. In addition, the vortex structure generated by pyrolysis also increases heat transfer.

It is worth noting that coking is often accompanied by pyrolysis, so it is necessary to further consider coking in pyrolysis-coupled heat transfer. An experiment was conducted by Wang et al. [23] to study the impact of the S-bend tube and mass flow rate on the thermal cracking and coke deposition behavior of supercritical aviation kerosene RP-3. When the outlet temperature of RP-3 was 600–700 °C, the conversion rate of the S-bend tube was 9.3% higher than that of the straight tube, and the gas yield was also slightly increased. This can be due to the secondary flow and Dean vortex formed by the bending structure affecting the flow field structure. The dual effects of high temperature and low speed promote thermal cracking in the core area of the S-bend tube. While the secondary flow enhances the fuel mixing effect, it also increases the local resistance, making the coke precursor easier to deposit near the wall and causing additional coking. On the other hand, the increase in mass flow affects the fuel's temperature, velocity, and turbulent kinetic energy, thereby reducing the cracking conversion rate and coke deposition rate. Furthermore, Zhang et al. [24] proposed a framework for 2D dynamic coking research by simultaneously coupling the detailed pyrolysis model of n-decane that involved secondary reactions with the MC-II coking model. Coking kinetics and their effect on flow and heat transfer characteristics have been studied in depth. The results showed that the framework established by dynamic mesh technology can better simulate the coking process and reflect the uneven distribution of deposited coke. Maximum conversion of n-decane obtained simultaneously was independent of the coking process. Finally, three penalties for coking are summarized. The increase in the maximum temperature increases the risk of solid structure breaking, the increase in thermal resistance decreases cooling performance, and the decrease in flow cross-sectional area increases pressure drop.

3.2. *Combustion and Heat Transfer Coupling in the Combustion Chamber*

Due to the dual role of fuel in regenerative cooling technology, there is a highly coupled relationship between fuel pyrolysis and combustion, which determines the heat flux, wall temperature, and structural stress response of the combustion chamber wall, and is critical to the design of thermal protection systems. To study this coupling relationship, establishing a one-dimensional mathematical model with fast calculation is a relatively easy solution. A typical one-dimensional model contains three modules of the heat transfer process, as shown in Figure 2. Zhang et al. [25] evaluated the performance of active thermal protection systems at different Mach numbers and equivalence ratios by the above model. In addition, the performance of the active thermal protection system and the active–passive combined thermal protection system is compared and analyzed. The results show that the combined thermal protection system can effectively reduce the wall temperature of the combustion chamber due to the role of the passive layer insulation material.

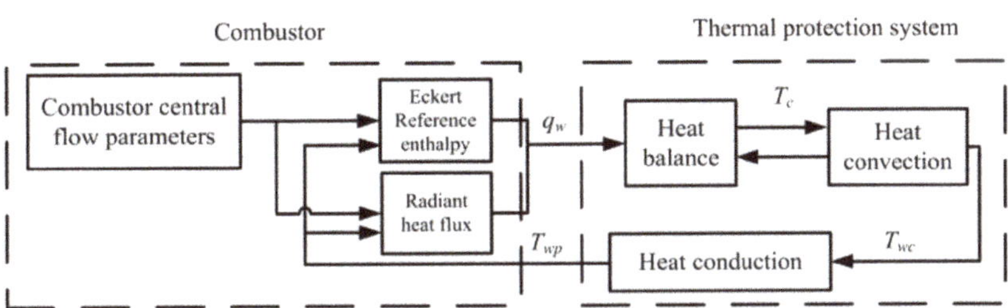

Figure 2. Schematic diagram of one-dimensional coupled heat transfer model of hydrogen fuel scramjet [25].

However, experimental studies can more realistically reflect this coupling relationship. Taddeo et al. [26] designed a regeneratively cooled combustor to study pyrolysis-combustion coupling and fuel-coking activity. Under the experimental conditions of ethylene as fuel and air as oxidant, the effects of fuel mass flow and equivalence ratio as two command parameters on the thermal environment of the combustion chamber were determined. The characterization parameters included the heat flux between the burned gas and the inner wall of the combustion chamber, the heat flux absorbed by the fuel in the cooling channel, and the heat transfer efficiency of the combustion chamber. The experimental results showed that when the first command parameters increased by 16–20%, the heat flux density of the corresponding gas side increased by 20% and 28%, which was determined by the equivalence ratio and pressure. The cooling side's physical and chemical heat fluxes were positively correlated with both command parameters. Similarly, the heat transfer efficiency of the combustor also increased with the increase in the fuel mass flow rate and changed nonlinearly with the increasing equivalence ratio, which means that there exists a maximum value. In addition, the sensitivity analysis of two characteristic heat fluxes to command parameters was performed. In the study of combustion conditions, it is found that when the equivalence ratio is greater than 1, radiative energy transfer accounts for more than thermal convection. The opposite is true when the equivalence ratio is equal to 1. Finally, the coking activity of ethylene was studied by monitoring the fuel pressure drop in the cooling channel.

In addition, the thermo-fluid-solid coupling is noteworthy. Gopinath et al. [27] numerically studied the three-dimension semi-couple transient fluid–thermal–structural response of the metallic sandwich panel with a corrugated core during hypersonic accelerated cruise-flight. The semi-coupled fluid–thermo–structural analysis methodology numerical framework is shown in Figure 3. The thermo-structural loads are estimated by adopting a high-speed gas-dynamic flow model bonded with Eckert's reference temperature. Taking the flight condition of Mach number 7 as an example, the coolant's flow and heat transfer characteristics are discussed, and the results of the directional displacement response of sandwich plates with and without cooling are compared and analyzed. The results show that the active cooling system not only improves the thermal protection capability by 70% but also reduces the bending deformation of the panel thermal structure response by 90%. In conclusion, the defect in the coupling of the thermal load and the cooling side is not considered in the method developed in this paper. Therefore, it is necessary to further develop the research method of flow–thermal–structural bidirectional coupling. To solve the problem of limited coolant for scramjet engines, Zhang et al. [28] proposed a topology optimization design method for the fluid-structure correlation of regenerative cooling channels to improve the overall heat transfer efficiency. The topology optimization structure of the cooling channel adopts the variable density method and takes heat transfer efficiency as the objective function. Under different heat flux distributions, the flow and heat transfer processes in the structure are quantitatively analyzed. The results show that the topological channel's average wall temperature and pressure drop are reduced by 5.8% and 20.6%, respectively, and the uniformity of the hot wall temperature is increased twice. The reason is that the microchannel perpendicular to the mainstream direction can distribute the inlet fuel to the adiabatic wall surface, thereby improving flow uniformity. However, the backflow phenomenon at the inlet and the "capillary" zone near the outlet will weaken the local heat transfer, resulting in a local high-temperature zone that needs further study.

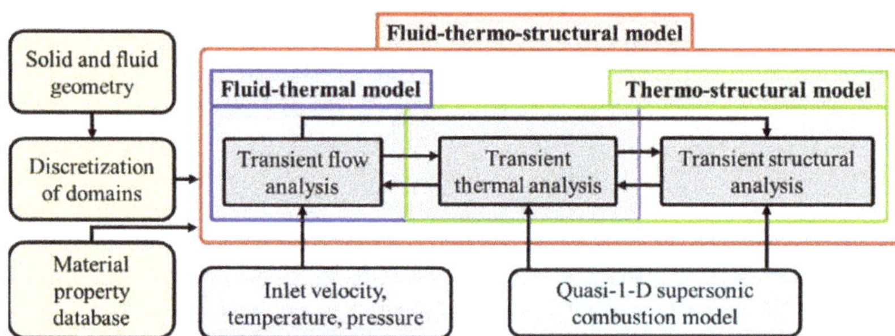

Figure 3. Numerical framework of the semi-coupled fluid–thermo–structural analysis methodology [27].

3.3. Conjugate Thermal Analysis Considering Aerodynamic Heating

To evaluate the thermal state of the components of a scramjet engine, it is necessary to comprehensively consider aerodynamic heating, combustion heat release, and cooling heat transfer, which require conjugate thermal analysis. Han et al. [10] innovatively developed a method for conjugate thermal analysis of regeneratively cooled scramjet engines. The scheme uses the principle of energy conservation to iteratively calculate the aerodynamic heating of the aircraft's inner and outer surface, the internal combustion heat release and structural thermal analysis, and the convective heat transfer of the coolant on the two interfaces. Specifically, the external flow field of the aircraft was calculated under adiabatic and isothermal boundary conditions. The aerodynamic heating of the scramjet surface was calculated by the direct correction method and used as the input condition to solve the structural thermal equation. Moreover, the additional heat load caused by the combustion phenomenon was reflected by increasing the adiabatic surface temperature in the combustion chamber. In addition, based on the assumption that the internal space of the aircraft is full of fuel and saturated steam, the average temperature of its volume was estimated. Concomitantly, the flow rate of n-dodecane in the cooling channel was calculated. The schematic diagram of the thermal analysis scheme and the flowchart of conjugate heat transfer is shown in Figures 4 and 5, respectively. Thus, the temperature fields of the outer surface of the scramjet, the inner surface near the combustion chamber, the internal space, and the cooling channel were obtained. Comparing the numerical results for Mach 6 at an altitude of 22.342 km, it was found that the coolant outlet temperature increased by up to 107 K after considering the aerodynamic heating absorbed by the internal space from the thermal structure. In some exceptional cases, however, cooling channels further heat the internal space in a complex way.

Furthermore, based on the developed numerical calculation method, the appropriate material can be selected subsequently according to the temperature of each component, which can also be used as the basis for the relevant design parameters of the cooling channel, such as the mass flow rate of the coolant, the inlet temperature and pressure, and the geometric parameters of the channel. In addition, an appropriate arrangement of payloads consisting of the internal system from a thermal point of view can be designed based on the internal space temperature of the aircraft. In summary, the advantage of conjugate thermal analysis is the ability to consider the coupling of all factors, which can be an important reference for the optimal design of the thermal protection system of the scramjet engine.

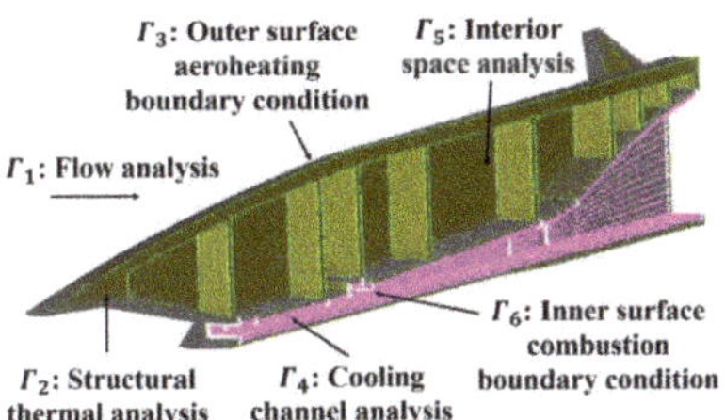

Figure 4. Regimes considered for thermal analysis of X-51A-like aircraft [10]. (The purple marking region in the figure represents the internal flow region of the engine including the cooling channel).

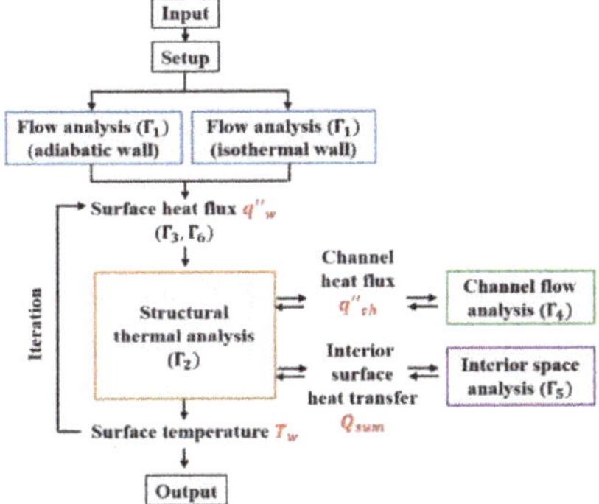

Figure 5. Flowchart of conjugate heat transfer [10].

4. Conclusions

This paper illustrates the necessity of research on coupled heat transfer of regenerative cooling technology for scramjet engines and summarizes its research progress. The following conclusions can be drawn:

(1) Pyrolysis in the cooling channel is not always conducive to heat transfer but is related to the heat flux and coking rate. Therefore, it is worthy of further study to regulate the cooling channel layout according to the actual heat flow distribution and suppress coking;

(2) The coupling relationship between combustion and heat transfer in the combustion chamber needs more in-depth research to achieve the optimal collaborative design of cooling performance and thrust performance. On the other hand, apart from calculating the one-way thermal structure response, the two-way coupling of the structure after thermal deformation and the flow field also needs to be explored;

(3) Further validating and improving the accuracy of the conjugate thermal analysis model and using it to evaluate the performance of various thermal protection systems can significantly save test and time costs.

Author Contributions: Conceptualization, C.L.; writing—original draft preparation, N.H.; writing—review and editing, Y.P., C.L. and J.L. All authors have read and agreed to the published version of the manuscript.

Funding: This research received no external funding.

Data Availability Statement: Not applicable.

Conflicts of Interest: The authors declare no conflict of interest.

References

1. Fry, R.S. A century of ramjet propulsion technology evolution. *J. Propul. Power* **2004**, *20*, 27–58. [CrossRef]
2. Mosavat, M.; Moradi, R.; Barzegar, M. The influence of coolant jet direction on heat reduction on the nose cone with Aerodome at supersonic flow. *J. Acta Astronaut.* **2018**, *6*, 26.

3. Curran, E.T. Scramjet engines: The first forty years. *J. Propul. Power* **2001**, *17*, 1138–1148. [CrossRef]
4. Barzegar, M.; Amini, Y.; Fallah, K. The flow feature of transverse hydrogen jet in presence of micro air jets in supersonic flow. *J. Adv. Space Res.* **2016**, *11*, 40.
5. Song, K.D.; Choi, S.H.; Scotti, S.J. Transpiration cooling experiment for scramjet engine combustion chamber by high heat fluxes. *J. Propul. Power* **2006**, *22*, 96–102. [CrossRef]
6. Hank, J.; Murphy, A.; Mutzman, R. The X-51A scramjet engine flight demonstration program. In Proceedings of the AIAA International Space Planes & Hypersonic Systems & Technologies Conference, Dayton, OH, USA, 28 April–1 May 2008.
7. Gascoin, N.; Abraham, G.; Gillard, P. Thermal and hydraulic effects of coke deposit in the hydrocarbon pyrolysis process. *J. Thermophys. Heat Transf.* **2012**, *26*, 57–65. [CrossRef]
8. Miao, H.Y.; Wang, Z.W.; Niu, Y.B. Performance Analysis of Cooling System Based on Improved Supercritical CO2 Brayton Cycle for Scramjet. *J. Appl. Therm. Eng.* **2019**, *19*, 114774. [CrossRef]
9. Zhao, Y.; Wang, Y.; Liang, C. Heat transfer analysis of n-decane with variable heat flux distributions in a mini-channel. *Appl. Therm. Eng.* **2018**, *144*, 695–701. [CrossRef]
10. Han, D.B.; Kim, J.S.; Kim, K.H. Conjugate thermal analysis of X-51A-like aircraft with regenerative cooling channels. *Aerosp. Sci. Technol.* **2022**, *126*, 107614. [CrossRef]
11. Tian, K.; Yang, P.; Tang, Z.C. Effect of pyrolytic reaction of supercritical aviation kerosene RP-3 on heat and mass transfer in the near-wall region. *Appl. Therm. Eng.* **2021**, *197*, 117401. [CrossRef]
12. Feng, T.; Qin, J.; Zhang, S.L. Modeling and analysis of heat and mass transfers of a supercritical hydrocarbon fuel with pyrolysis in mini-channel. *J. Heat Transfer* **2015**, *91*, 520–531. [CrossRef]
13. Feng, T.; Zhang, S.L.; Cao, J. Coupling relationship analysis between flow and pyrolysis reaction of endothermic hydrocarbon fuel given characteristic time correlation in mini-channel. *Appl. Therm. Eng.* **2016**, *16*, 30486.
14. Jiang, Y.G.; Qin, J.; Chetehouna, K. Effect of geometry parameters on the hydrocarbon fuel flow rate distribution in pyrolysis zone of SCRamjet cooling channels. *J. Heat Transf.* **2019**, *141*, 1114–1130. [CrossRef]
15. Lei, Z.L.; He, K.; Huang, Q. Numerical study on supercritical heat transfer of n-decane during pyrolysis in rectangular tubes. *Appl. Therm. Eng.* **2020**, *170*, 115002. [CrossRef]
16. Tian, K.; Tang, Z.C.; Wang, J. Simplified pyrolytic model of RP-3 fuel at high conversion rates and the effects of primary reaction on the regenerative cooling process. *Aerosp. Sci. Technol.* **2022**, *129*, 107853. [CrossRef]
17. Li, Z.Z.; Li, Y.; Zhang, X.W. Coupling of pyrolysis and heat transfer of supercritical hydrocarbon fuel in rectangular minichannels. *Chem. Eng. Sci.* **2022**, *247*, 116924. [CrossRef]
18. Li, Z.Z.; Liu, G.Z.; Zhang, R.L. Heat transfer to supercritical hydrocarbon fuel in horizontal tube: Effects of near-wall pyrolysis at high heat flux. *Chem. Eng. Sci.* **2021**, *229*, 115994. [CrossRef]
19. Gong, K.Y.; Cao, Y.; Feng, Y. Influence of secondary reactions on heat transfer process during pyrolysis of hydrocarbon fuel under supercritical conditions. *Appl. Therm. Eng.* **2019**, *159*, 113912. [CrossRef]
20. Jiao, S.; Li, S.F.; Pu, H. Experimental investigation on thermal cracking and convective heat transfer characteristics of aviation kerosene RP-3 in a vertical tube under supercritical pressures. *J. Therm. Sci.* **2019**, *146*, 106092. [CrossRef]
21. Jiao, S.; Li, S.F.; Pu, H. Investigation of pyrolysis effect on convective heat transfer characteristics of supercritical aviation kerosene. *Acta Astronaut.* **2020**, *171*, 55–68. [CrossRef]
22. Li, Y.; Wang, J.B.; Xie, G.N. Effect of thermal pyrolysis on heat transfer and upward flow characteristics in a rectangular channel using endothermic hydrocarbon fuel. *Chem. Eng. Sci.* **2021**, *244*, 116806. [CrossRef]
23. Wang, R.T.; Zhao, J.J.; Bao, Z.W. Thermal cracking and coke deposition characteristics of aviation kerosene RP-3 in an S-bend tube. *Fuel* **2022**, *313*, 122673. [CrossRef]
24. Zhang, C.; Gao, H.; Zhao, J.J. Dynamic coking simulation of supercritical n-decane in circular tubes. *Fuel* **2023**, *331*, 125859. [CrossRef]
25. Zhang, C.; Qin, J.; Yang, Q.C. Design and heat transfer characteristics analysis of combined active and passive thermal protection system for hydrogen-fueled scramjet. *Int. J. Hydrogen Energy* **2015**, *40*, 675–682. [CrossRef]
26. Taddeo, L.; Gascoin, N.; Chetehouna, K. Experimental study of pyrolysis-combustion coupling in a regeneratively cooled combustor: Heat transfer and coke formation. *Fuel* **2019**, *239*, 1091–1101. [CrossRef]
27. Gopinath, N.K.; Govindarajan, K.V.; Mahapatra, D.R. Fluid-thermo-structural response of actively cooled scramjet combustor in hypersonic accelerating-cruise flight. *J. Heat Transf.* **2022**, *194*, 123060. [CrossRef]
28. Zhang, T.; Jing, T.T.; Qin, J. Topology optimization of the regenerative cooling channel in the non-uniform thermal environment of hypersonic engineopology optimization of regenerative cooli thermal environment of the hypersonic engine. *Appl. Therm. Eng.* **2023**, *219*, 119384. [CrossRef]

Article

Numerical Study of the Effect of the Rolling Motion on the Subcooled Flow Boiling in the Subchannel

Yaru Li, Xiangyu Chi, Zezhao Nan, Xuan Yin, Xiaohan Ren and Naihua Wang *

Institute of Thermal Science and Technology, Shandong University, Jinan 250061, China; lyr1993@yeah.net (Y.L.); chi2072478@126.com (X.C.); sdnanzezhao@163.com (Z.N.); xuan_yin1214@163.com (X.Y.); xiaohan09126@gmail.com (X.R.)
* Correspondence: wnh@sdu.edu.cn

Abstract: The marine environment may change the force on the fluid and inevitably influence bubble behavior and the two-phase flow in the reactor core, which are vital to the safety margin of a nuclear reactor. To explore the effect of the marine motion on the flow and heat transfer features of subcooled flow boiling in the reactor core, the volume of fluid (VOF) method is employed to reveal the interaction between the interface structure and two-phase flow in the subchannel under rolling motion. The variations of several physical parameters are obtained, including the transverse flow, the vapor volume fraction, the vapor adhesion ratio, and the phase distribution of boiling two-phase flow with time. Sensitivity analyses of the amplitude and the period of the rolling motion were performed to demonstrate the mechanisms of the influence of the rolling motion. We found that the transverse flow in the subchannel was mainly affected by the Euler force under the rolling motion. In contrast to the two-phase flow in the static state, the vapor volume fraction and vapor adhesion ratio show different characteristics under rolling motion. Additionally, the onset of significant void (OSV) point changes periodically under rolling motion.

Keywords: rolling motion; subcooled flow boiling; transverse flow; subchannel; VOF

Citation: Li, Y.; Chi, X.; Nan, Z.; Yin, X.; Ren, X.; Wang, N. Numerical Study of the Effect of the Rolling Motion on the Subcooled Flow Boiling in the Subchannel. *Energies* **2022**, *15*, 4866. https://doi.org/10.3390/en15134866

Academic Editors: Jian Liu, Bengt Sunden, Chaoyang Liu and Yiheng Tong

Received: 15 June 2022
Accepted: 30 June 2022
Published: 2 July 2022

1. Introduction

An offshore floating nuclear plant combines a nuclear power plant and a floating platform. Its flexibility is a huge advantage in energy supply. However, marine conditions will change the force on the fluid and inevitably influence the two-phase flow, bubble behavior, and void fraction in the reactor core, which are vital to the occurrence of boiling crisis phenomena and the safety margin of the reactor core.

Various researchers have experimentally studied two-phase flow under rolling motion. Yan et al. [1] investigated the interfacial parameters of an adiabatic bubbly flow in a narrow rectangular channel under rolling conditions. The test section offsetting the rolling axis was installed on the rolling platform. The rolling parameters were amplitudes from 10° to 30° and periods from 8 s to 16 s. There were no obvious differences among interfacial parameters at the same position under inclined and rolling conditions. They deduced that this is because the buoyancy induced by the rolling motion was far less than the lateral component of the buoyancy induced by gravity. Tanjung et al. [2] experimentally investigated the bubble behavior in the saturated pool boiling under rolling motion. The test section was installed above and below the rolling axis. The rolling parameters were amplitudes of 10° and 30° and periods of 10 s and 20 s. They found that the tangential and centrifugal inertia force in the non-inertia frame significantly affected the bubble behaviors. These forces had an opposite effect on the bubble behaviors when the relative positions of the test section and the rolling axis were different. Hwang et al. [3] carried out an experiment on the subcooled flow boiling in a circular tube. The test loop was installed below the rolling axis. The rolling parameters were amplitudes from 15° to 40° and periods from 6 s to 18 s. The effect of the rolling motion on the critical heat flux was weak when the period was above 6 s. Compared with the static state, the critical heat flux under the rolling motion was higher than the static state for the high pressure of 24 bar, while it was affected by the mass flux and the rolling amplitude for the low pressure. Li et al. [4,5] conducted a visualization experiment to study bubble behaviors of a subcooled flow boiling under rolling motion. The test section was a narrow rectangular channel with a single-heated surface. The test loop offsetting the rolling axis was installed on the rolling platform. The rolling parameters were amplitudes from 8° to 16° and periods of 10 s to 15 s. The effects of additional force, variations in local pressure, and flow rate oscillation on bubble behavior were analyzed. They found that the rolling motion changed the parameters of bubbles due to the variation in the local pressure drop.

The conclusions of the above studies do not coincide or even contradict each other on the effect of rolling motion for the following three reasons. Firstly, the working conditions and the structure of the test section are widely varied. These factors (the adiabatic vs. boiling two-phase flow, or the narrow rectangular vs. circular channel) have different effects on the flow and characteristics, not to mention the effect of the rolling motion. Secondly, the fictitious force on the bubble varied with the positions of the rolling axis and the test section and may have an opposite effect on the bubble behavior and trajectory. Thirdly, the effect of the rolling motion on the test section was hardly separated from the effects of pressure or mass variations arising from the test loop, which is also under the rolling motion.

Experimental research is difficult to conduct for extremely detailed phenomena such as bubble behaviors. Recently, the investigation of two-phase flow by numerical methods was developed with the improved multiphase flow model and affordable computing power. The CFD techniques can provide detailed spatial and temporal distributions of various parameters [6–8]. The numerical methods for two-phase flow can be divided into two branches according to whether the interface between immiscible fluids is tracked or not. In the present study, the vapor and liquid are well separated for the flow boiling in the reactor core. Moreover, the interface structure plays a key role in the two-phase flow and heat transfer, which may differ from the existing literature conclusions due to the coupling of the phase change and the marine motion [9]. Consequently, the interface-tracking method is the only feasible and viable choice for this research. The VOF method has a clear advantage of intrinsic mass conservation among the typical interface-tracking methods. The numerical simulation of the flow boiling with high inlet subcooling in the reactor core can be accomplished with a combination of the mass transfer model called the Lee model.

The behavior of isolated bubbles in the pool boiling [10,11] and flow boiling has been investigated using the VOF model. Only the natural convective flow needs to be considered in the former condition. The deformation and trajectory of an isolated bubble in the latter condition are more complicated, owing to the interaction between the bubble and the forced flow. Bahreini et al. [12] and Jeon et al. [13] simulated the subcooled flow boiling with a uniform and a linear inlet velocity profile. Bahreini et al. suggested that the bubble shape was different between the two cases because the inertial force of the surrounding liquid changed. Jeon et al. found that the condensing bubble finally changed into an ellipsoidal shape while the adiabatic bubble changed into a wing shape. Guo et al. [14] and Cheng et al. [15] simulated the growth of a single bubble in the subcooled flow boiling and investigated the effect of the microlayer during the bubble ebullition. The microlayer contributed greatly to the increase in the heat transfer coefficient, and 30% of the evaporation occurred in the microlayer. Pattamatta et al. [16] and Liu et al. [17] investigated the interaction of two and three bubbles in a rectangular channel. Pattamatta et al. found that the coalescence time was shorter when the downstream bubble was smaller than the upper one. In the study of Liu et al., the random perturbation of the upper bubble would accelerate the condensation of the lower bubble.

In the actual flow boiling process, the growth process and the interaction of bubbles are quite different from those of isolated bubbles, which essentially affect the two-phase flow regime and the heat transfer pattern. Lee et al. [18] simulated the subcooled flow boiling in a rectangular channel with two opposite heated surfaces. The bubble was less spherical for a larger inlet velocity. In addition, the large bubble broke into small bubbles in the downstream region due to the shear stress induced by larger velocity. Kuang et al. [19] and Yin et al. [20] investigated the flow boiling in an intermediate-diameter tube under low heat and mass flux. The slug bubbles were unstable in the upper section and likely broke into churn slugs or churn froth. They suggested that the heat transfer coefficient was dominated by the nucleate boiling under the given geometry and working conditions. Devahdhanush et al. [21] compared two-phase interfacial behavior for the flow boiling in macro- and micro-channels. The heat transfer performance of micro-channels was better than that of macro-channels. However, bubbles in the micro-channel tended to grow in the radial direction and occupied the entire flow area. This increased the possibility of two-phase choking. Darshan et al. [22] found that the boiling performance in the dimpled tube is better than that in the plain tube because the cavities on the heated surface promoted the initiation of bubble nucleation and contributed to the increase in nucleation sites. In addition, the departure frequency in the dimpled tube was larger than that in the plain tube. Zhuan et al. [23] and Yang et al. [24] studied the flow boiling in the horizontal straight and coiled tube. Zhuan et al. claimed that the bubble behavior significantly affected the transition of two-phase flow patterns. Yang et al. pointed out that the physical field in the tube bend differed from that in the straight section under the effects of buoyance and centrifugal force.

Several studies focused on the subcooled flow boiling under the marine environment using the VOF model. Amidu et al. [25] investigated the subcooled flow boiling in an inclined rectangular channel with a single-heated surface facing downward. The bubble was attached to the heated surface due to the buoyancy effect. Then, it slid along the surface under the drag force induced by the surrounding liquid. Bahreini et al. [26] and Lee et al. [27] studied the flow boiling in a rectangular channel under microgravity conditions. In their study, larger bubbles formed in the microgravity case than in the normal gravity case. However, the variation in bubble size had a contrary effect on the heat transfer. Bahreini et al. indicated that lateral coalescence of bubbles enhanced the heat transfer. On the contrary, Lee found larger bubbles coalesced into the vapor blanket, and the heat transfer deteriorated. Wang et al. [28] simulated the pool film boiling at a horizontal plate heater under different gravity magnitudes and inclination angles. The result showed that bubble departure diameter increased with the decreasing gravity. The heat flux increased with the inclination angle, and this was because bubbles moved along the heated surface and disturbed the flow. Wei et al. [29] compared the subcooled flow boiling in the static and swing cases. They concluded that the bubble shape was similar while the bubble distribution was different because of the variation in mass flow rate.

In summary, the coupling of multiple factors has a complex effect on bubble behaviors. Firstly, the hydrodynamics vary with the structure of the channel. Secondly, the bubble structure and the mechanism of the boiling crisis are strongly related to the working conditions. Thirdly, in addition to the orientation of the heated surface, the transient body force induced by marine motion also affects the two-phase flow in the heated channel. All these factors must be satisfied to investigate the effect of marine motion on the two-phase flow in the reactor core. However, little research has been carried out in this regard.

The main purpose of this study was to analyze the effect and mechanism of the rolling motion on the boiling two-phase flow in the subchannel. We established a numerical model to simulate the flow boiling with high heat flux and inlet subcooling in the subchannel by the VOF approach. This model was validated for the static case, and a mesh independence study was performed. In the present study, the tendency of the transient physical parameters and distributions, including the transverse velocity, the vapor volume fraction, the vapor adhesion ratio, and the phase distribution, are described along the heated channel. The cause and effect between these physical parameters and the rolling motion are discussed. Moreover, a sensitivity analysis was carried out to demonstrate how the rolling motion amplitude and period affect the bubble behavior of subcooled flow boiling in the subchannel.

2. Computational Methods

2.1. Governing Equations

The capability of the VOF model with the Lee model to predict the subcooled flow boiling has been demonstrated in the published literature. In this study, a three-dimensional computational model is established to simulate the two-phase flow in the subchannel by combining the VOF model and the Lee model. In the VOF method, the liquid–vapor interface can be tracked transiently. Liquid and vapor are considered as individually immiscible and incompressible phases. In each control volume, the volume fractions of liquid and vapor phases sum to unity, and only the latter α_v is solved.

$$\frac{\partial}{\partial t}(\alpha_v\rho_v) + \nabla \cdot \left(\alpha_v\rho_v\vec{v}\right) = \dot{m}_{lv} - \dot{m}_{vl} \tag{1}$$

The mass source term on the right-hand side of Equation (1) represents the mass transfer due to the phase change. The interface is calculated by the geometric reconstruction scheme. The velocity field solved by the momentum equation, Equation (2), is shared by both the liquid and vapor phases.

$$\frac{\partial}{\partial t}\left(\rho\vec{v}\right) + \nabla \cdot \left(\rho\vec{v}\vec{v}\right) = -\nabla p + \nabla \cdot \left[\mu\left(\nabla\vec{v} + \nabla\vec{v}^{T}\right)\right] + \rho\vec{g} + \vec{F_s} \tag{2}$$

The surface tension force $\vec{F_s}$ is modeled by the continuum surface force (CSF) model proposed by Brackbill et al. [30], in which the wall adhesion effect is considered. In the present study, the surface tension coefficient and the contact angle are 5.4943×10^{-3} and 90°, respectively. The energy equation is given by

$$\frac{\partial}{\partial t}(\rho E) + \nabla \cdot \left(\vec{v}(\rho E + p)\right) = \nabla \cdot (k\nabla T'') + S_h, \tag{3}$$

where E is a mass-averaged variable:

$$E = \frac{\alpha_v\rho_v E_v + \alpha_l\rho_l E_l}{\alpha_v\rho_v + \alpha_l\rho_l}. \tag{4}$$

The energy for vapor phase, E_v, is given by

$$E_v = c_{p,v}T'' - \frac{p}{\rho_v} + \frac{v^2}{2}. \tag{5}$$

The energy for liquid phase, E_l, is given by

$$E_l = c_{p,l}T'' - \frac{p}{\rho_l} + \frac{v^2}{2}. \tag{6}$$

For a computation cell full of liquid or vapor, the material properties of the working fluid are piecewise linear functions of temperature, which are obtained from the NIST Chemistry WebBook website [31]. For a computation cell that contains both liquid and vapor, the material properties are volume fraction-averaged values. For instance, the density is represented as

$$\rho = \alpha_v\rho_v + (1 - \alpha_v)\rho_l. \tag{7}$$

The phase change process is accomplished by the Lee model. The difference between the local and saturated temperatures determines whether the phase change occurs. Therefore, the nucleation can be simulated in the saturated phase without pre-existing interfaces. In the Lee model, the mass transfer rate depends on the below temperature regime.

$$\dot{m}_{\text{lv}} = \beta_{\text{l}}\alpha_{\text{l}}\rho_{\text{l}}\frac{\left(T''_{\text{l}} - T''_{\text{sat}}\right)}{T''_{\text{sat}}} \tag{8}$$

$$\dot{m}_{\text{vl}} = \beta_{\text{v}}\alpha_{\text{v}}\rho_{\text{v}}\frac{\left(T''_{\text{sat}} - T''_{\text{v}}\right)}{T''_{\text{sat}}} \tag{9}$$

In the current work, the empirical coefficients β_{l} and β_{v} are set as 160 s^{-1} and 100 s^{-1}, respectively. The heat transfer rate is obtained by multiplying the mass transfer rate by the latent heat, a constant of 1019.4 kJ/kg.

The realizable k-ε turbulence model with the standard wall function is adopted in this simulation, considering the high Reynolds number flow and secondary flow in the subchannel. The mesh generation uses the ICEM CFD software. The results of four meshes with different cell sizes in the near-wall region are compared with the experimental result. The mesh with a total number of 8.30 million is adopted.

The numerical calculation is implemented by the Fluent v18.2 software. The pressure term is interpolated by the PRESTO! scheme. The pressure–velocity coupling is treated with the PISO scheme. The least square cell-based method is used for gradient discretization. The numerical convergence has been performed in a prior study.

2.2. Computational Domain and Boundary Conditions

Coolant flows through the subchannels formed between the fuel rods to remove heat from the reactor core. The rod bundles can be subdivided into rod-centered or coolant-centered subchannels. In the present study, a coolant-centered subchannel arranged in a square array is selected as the computational domain. The rod diameter and pitch distance are 9.5 mm and 3.1 mm, respectively. The heated length is 1.555 m. A rectangular region is connected to each rod gap, making the computational domain identical to the test section. This region is in the shape of 1.55 mm × 3.1 mm and is referred to as a connected area (Figure 1a). Given that the structure of the channel affects the hydrodynamic characteristics, we investigate the effect of the marine environment in isolation from the cross-flow between subchannels and the secondary flow due to the spacer grid.

The two-phase flow phenomena in different regions of the subchannel are discussed in Section 3. For convenience, these regions are defined as follows (Figure 1b,c). CS-H_1 represents the cross-section which is H_1 from the inlet. The flow domain between CS-H_1 and CS-H_2 is Domain H_1-H_2. It is divided into four quadrants numbered anticlockwise. The northeast one is Quadrant-1 (of Domain H_1-H_2). Segment H_1-H_2 includes all the heated surfaces and rectangular unheated surfaces adjacent to them in Domain H_1-H_2. The northeast one is Wall-1 (of Segment H_1-H_2).

The boundary conditions are set according to No.1.2211 of the subchannel test in NUPEC PSBT [32]. This working condition of this test is close to that of the normal operation of the pressurized water reactor (PWR). The test is performed under the pressure of 14.72 MPa. The heating power is 90 kW. The inlet temperature and the mass flux are 568.55 K and 3030.6 kg/$(\text{m}^2{\cdot}\text{s})$, respectively. The boundary conditions are shown in Table 1. Curved surfaces representing fuel rods are heated with consistent heat flux. The three surfaces around a connected area are unheated.

Table 1. Boundary conditions.

Position	Boundary Condition	Parameter and Value
Inlet	Velocity inlet	$\alpha_{\text{v}} = 0,\ v_x = v_y = 0,\ v_z = 4.13\ \text{m/s},\ T'' = 568.55\ \text{K}$
Heated wall	No-slip wall	$\frac{\partial \alpha_{\text{v}}}{\partial \vec{n}} = 0,\ v_x = v_y = v_z = 0,\ k\frac{\partial T''}{\partial \vec{n}} = 1939.3\ \text{KW/m}^2$
Unheated wall	No-slip wall	$\frac{\partial \alpha_{\text{v}}}{\partial \vec{n}} = 0,\ v_x = v_y = v_z = 0,\ k\frac{\partial T''}{\partial \vec{n}} = 0$

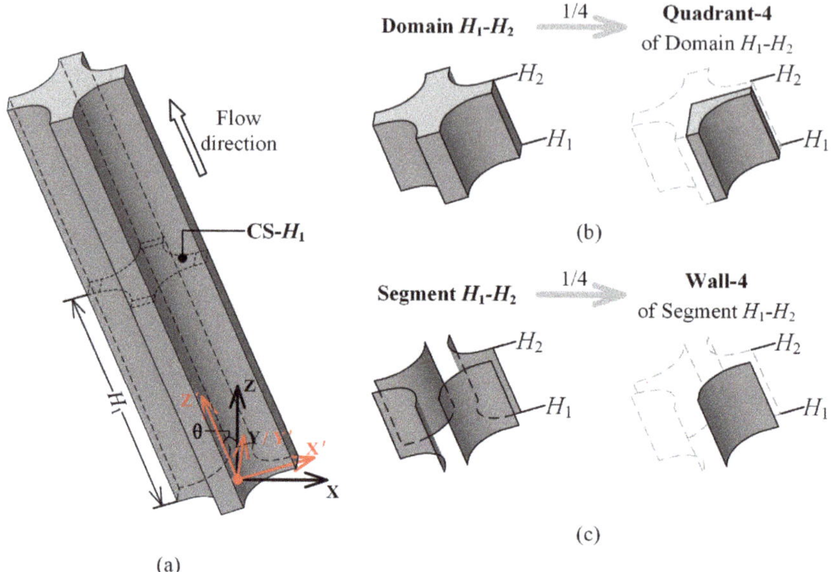

Figure 1. Computational domain. (**a**) Subchannel; (**b**) domain and quadrant; (**c**) segment and wall.

2.3. Marine Motion

The sliding mesh model (SMM) is used to model the subchannel moving under marine motion. In the SMM, the boundaries of the subchannel and cells in the flow domain move together in a rigid body motion. The shape of each cell is unchanged. The node of the cell is a function of time and is updated in each time step. Therefore, the governing equations are written as

$$\frac{\partial}{\partial t}(\alpha_{\rm v}\rho_{\rm v}) + \nabla\cdot\left(\alpha_{\rm v}\rho_{\rm v}\left(\vec{v}-\vec{v_{\rm m}}\right)\right) = \dot{m}_{\rm lv} - \dot{m}_{\rm vl}, \tag{10}$$

$$\frac{\partial}{\partial t}\left(\rho\vec{v}\right) + \nabla\cdot\left(\rho\vec{v}\left(\vec{v}-\vec{v_{\rm m}}\right)\right) = -\nabla p + \nabla\cdot\left[\mu\left(\nabla\vec{v}+\nabla\vec{v}^{T}\right)\right] + \rho\vec{g} + \vec{F_{\rm s}}, \tag{11}$$

$$\frac{\partial}{\partial t}(\rho E) + \nabla\cdot\left(\left(\vec{v}-\vec{v_{\rm m}}\right)(\rho E + p)\right) = \nabla\cdot\left(k\nabla T''\right) + S_{\rm h}, \tag{12}$$

where $\vec{v_{\rm m}}$ is the velocity of the moving boundary and the cells.

In the present study, the marine motion is simplified to the rolling motion following the trigonometric function. In the rolling motion, angle θ at time t is defined by

$$\theta(t) = \theta_{\rm m}\sin\left(\frac{2\pi}{T}t\right), \tag{13}$$

where $\theta_{\rm m}$ and T are the amplitude and the period, respectively. Then, the angular velocity ω and the angular acceleration ε can be given as

$$\omega(t) = \dot{\theta} = \frac{2\pi}{T}\theta_{\rm m}\cos\left(\frac{2\pi}{T}t\right), \tag{14}$$

$$\varepsilon(t) = \ddot{\theta} = -\frac{4\pi^2}{T^2}\theta_{\rm m}\sin\left(\frac{2\pi}{T}t\right). \tag{15}$$

The y-axis of the reference frame S is the rotational axis of the rolling motion. At the beginning of the rolling motion, the subchannel starts to rotate from the equilibrium position, i.e., the vertical position, to the extreme position in the negative direction of the x-axis.

A non-inertial reference frame S' with axes x', y', and z' is established for the convenience of presenting and analyzing the results. The origin is located at the center of the inlet (the red one in Figure 1). The non-inertial frame S' is attached to the subchannel under rolling motion. The y'-axis coincides with the y-axis. The z'-axis is parallel to the flow direction. Viewed from the frame S, the frame S' rotates with the angle θ. For a particle in the rotating frame S', Newton's second law is given by

$$m\vec{a}' = \vec{F} + \vec{F_\tau} + \vec{F_n} + \vec{F_c}. \quad (16)$$

Three further terms called fictitious forces are added to the right-hand side of Newton's law of motion. The three forces are the Euler force, the centrifugal force, and the Coriolis force,

$$\vec{F_\tau} = m\vec{a_\tau} = -m\vec{\varepsilon} \times \vec{r}, \quad (17)$$

$$\vec{F_n} = m\vec{a_n} = -m\vec{\omega} \times \left(\vec{\omega} \times \vec{r}\right), \quad (18)$$

$$\vec{F_c} = m\vec{a_c} = -2m\vec{\omega} \times \vec{v}', \quad (19)$$

where $\vec{v}'$ is the velocity of the particle in the frame S'.

2.4. Validation of the Numerical Model

We validated the numerical model by comparing the numerical prediction with the experimental data for the void fraction in the static case. The experiment performed by NUPEC consists of a series of void measurements using full-size mock-up tests for PWRs. Part of the experimental database has been made available for the OECD/NRC PSBT benchmark to validate numerical models. The working condition of run No.1.2211 of the experiment is closest to that of the normal operation of the PWR among the subchannel tests. Therefore, the experimental data of this run are selected for the validation of the numerical model. Additionally, the cross-sectional void fraction at a given elevation was the only data provided in the subchannel test.

A mesh independence study is performed using four different numbers of mesh cells. The numerical predictions for the cell numbers of 8 million and 10 million deviate from the experimental data by 11.9% and 10.4%, respectively (Figure 2). Given the complex two-phase flow behavior, the mesh with around 8 million cells is selected to provide accuracy with minimum computing time.

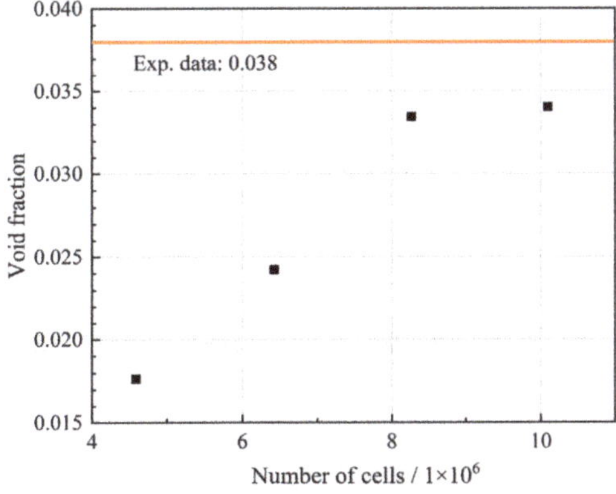

Figure 2. Numerical validation and mesh independence study.

3. Results and Discussion

The results of a rolling case and a static case are compared at first to discuss the two-phase flow features under rolling motion. It should be pointed out that the amplitude and the period of the rolling case mentioned above are 15° and 3 s, respectively. A sensitivity analysis of the amplitude and the period is carried out in Section 3.2 based on the discussion in Section 3.1.

In this section, the result of rolling cases is calculated in the non-inertial frame S' and presented within the time of a period. The time is in the range of 0 to T. A notable exception is that $t = -T/6$. It means the time that is $T/6$ before the beginning of the rolling motion in a period.

3.1. Effect of Rolling Motion

3.1.1. Transverse Flow

The secondary flow is classified into two kinds by the physical mechanism [33]. Secondary flows of the first kind, driven by pressure, occur in both laminar and turbulent flows. Secondary flows of the second kind due to the non-isotropic Reynolds stress distribution only occur in turbulent flow in straight pipes of a noncircular cross-section. In the present study, these two kinds of secondary flows both exist under the effect of cross-section, boiling, and rolling motion. This flow perpendicular to the main direction of fluid motion is referred to here as transverse flow. Figure 3 depicts the transverse velocities on the x-axis (x'-axis) of CS-1.10 m, CS-1.20 m, CS-1.30 m, and CS-1.40 m. The transverse velocities on the y-axis (y'-axis) are shown in Figure 4.

The transverse velocities in all the cross-sections are nearly constant with time under static state. The profiles of the transverse velocity on the x-axis and the y-axis are similar and feature two peaks. The transverse velocities are almost zero in the middle of the cross-section, whereas they reach the peaks in the rod gap. The transverse velocities in the rod gap increase with the height of the cross-section.

The transverse velocities under rolling motion differ significantly from those under static state. They change periodically. Firstly, transverse velocity profiles at the beginning of the period (Figures 3b and 4b) are almost the same as those at the end (Figures 3f and 4f). Secondly, the profiles of the transverse velocity on the x'-axis at the moments t and $t + T/2$ are symmetric with respect to the y'-axis. Thirdly, when the subchannel is in the equilibrium position, which is the same as that under static state, the transverse velocity under rolling motion is far more than that under static state, especially in the middle of the cross-section (Figure 3b,d).

The above-mentioned transverse flow is induced mainly by boiling and rolling motion. The liquid subcooling is too high to permit bubble nucleation near the subchannel inlet. The heat transfer regime is forced convection. With increasing height, boiling initiates when the liquid in the vicinity of the heated surface reaches saturated. Mass transfer between the liquid phase and the vapor phase and the disturbance of bubbles lead to transverse flow inside the subchannel.

Under static state, due to the symmetrical geometry of the subchannel, the transverse velocity fields at the cross-sections are axially symmetrical with two vortices rotating reversely in the rod gaps (Figure 5a), with a sketch of the subchannel in the lower left corner). A similar transverse flow field is also obtained in Bieder et al.'s study [34]. Further downstream, the nucleate boiling intensifies, and the transverse velocity increases. The nucleate boiling occurs only in the vicinity of the heated surface while the bulk liquid is still subcooled. As a result, there is almost no transverse flow at the cross-section center under static state. Moreover, flow and heat transfer in the subchannel are stable, and the transverse flow is independent of time.

Boiling will promote the transverse flow of fluid, whereas rolling motion has a predominant role under rolling motion. The formation and evolution of the transverse flow under rolling motion are interpreted with the transverse flow fields.

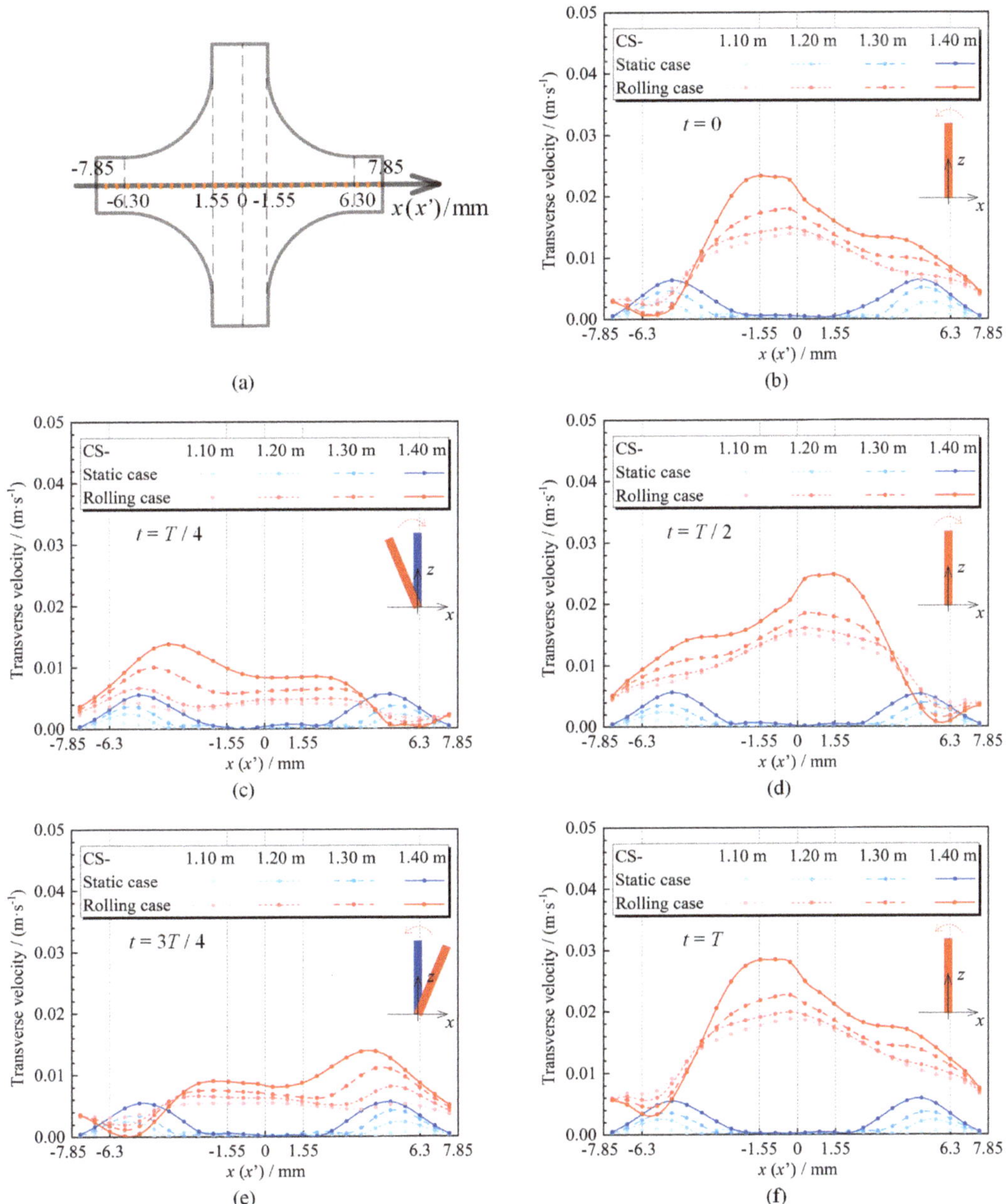

Figure 3. Transverse velocity profiles on the x-axis (x'-axis). (**a**) Cross-section of the subchannel; (**b**) $t = 0$; (**c**) $t = T/4$; (**d**) $t = T/2$; (**e**) $t = 3T/4$; (**f**) $t = T$.

When the subchannel rotates anticlockwise to the equilibrium position, the angular acceleration is negative (Figure 5b), and the Euler force on the fluid is positive along the x'-axis (Equation (11)). Meanwhile, a region consisting of the middle of the cross-section and the two gaps around the x'-axis is free of wall impediment. Fluid in this open region flows along the x'-axis to the positive direction under the action of the Euler force, and two secondary vortices of the reverse direction emerge at the cross-section (Figure 5c). On the negative side of the x'-axis, the transverse velocity induced by boiling and rolling motion is in the opposite direction. Therefore, the secondary vortex on the negative side of the x'-axis is weakened by the rolling motion and smaller than that under static state. On the positive side of the x'-axis, the secondary vortex induced by boiling merges into that induced by rolling motion because of the same direction.

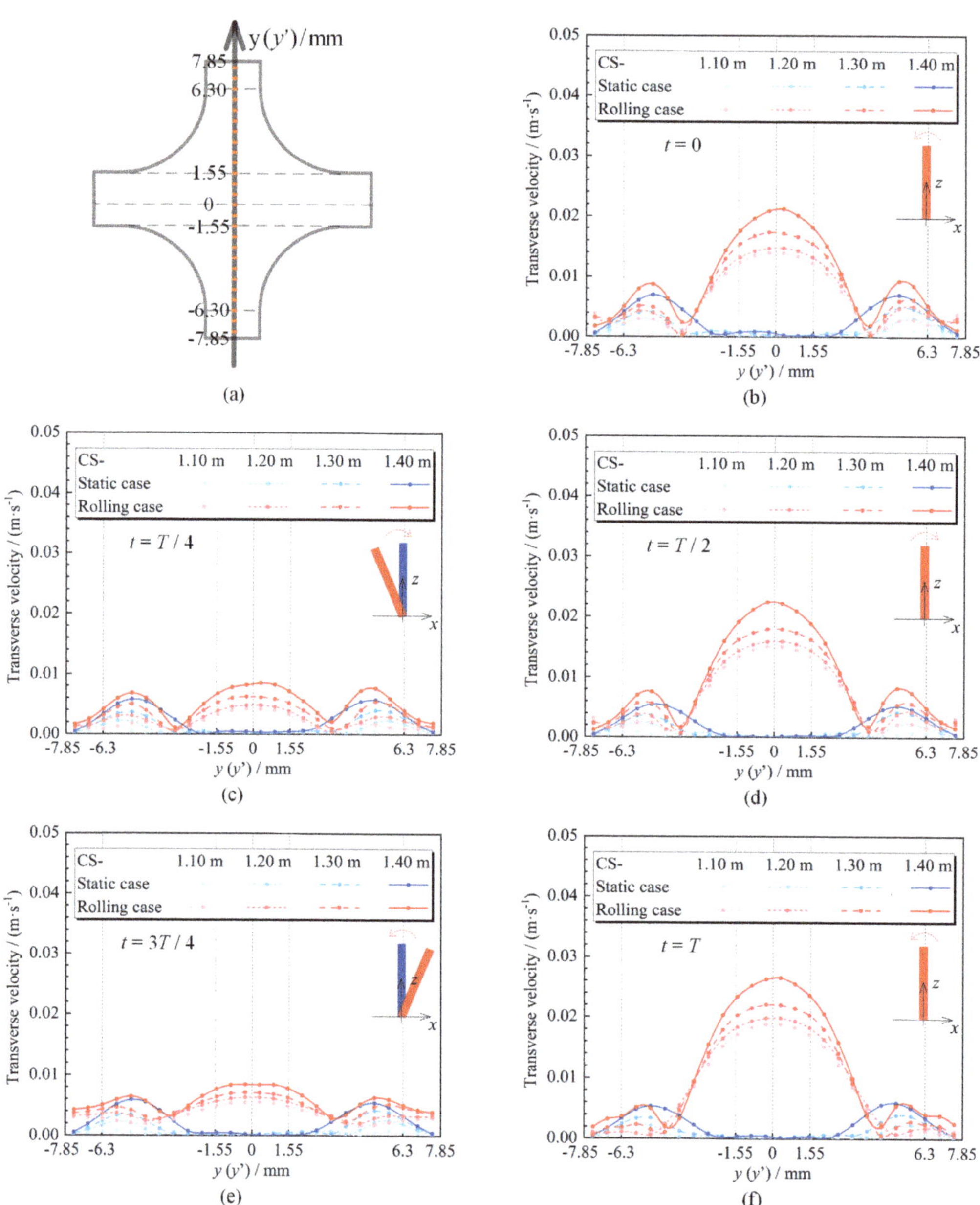

Figure 4. Transverse velocity profiles on the y-axis (y'-axis). (**a**) Cross-section of the subchannel; (**b**) $t = 0$; (**c**) $t = T/4$; (**d**) $t = T/2$; (**e**) $t = 3T/4$; (**f**) $t = T$.

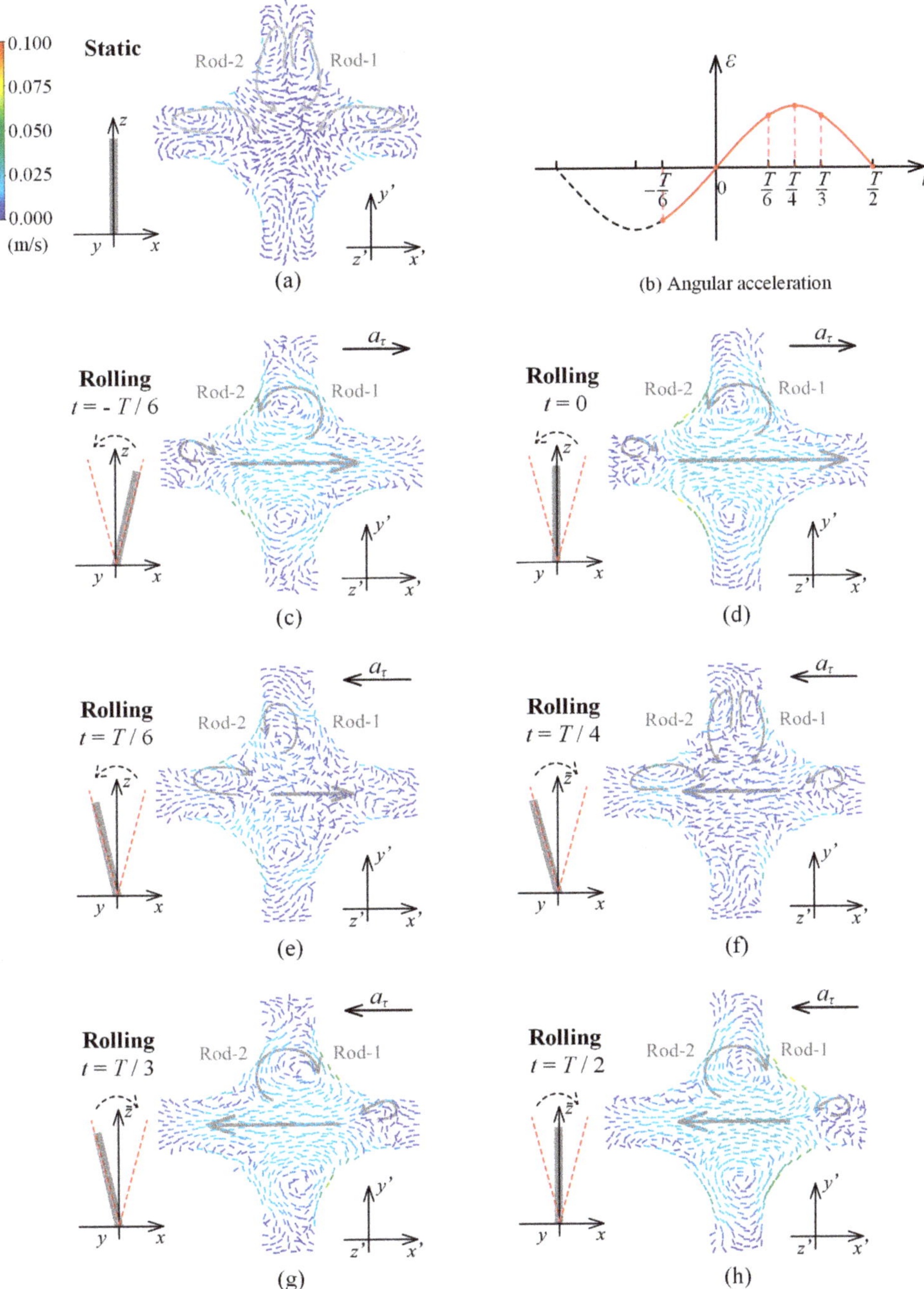

Figure 5. Transverse flow field. (**a**) Static case; (**b**) angular acceleration of the rolling case; (**c**) rolling case $t = -T/6$; (**d**) rolling case $t = 0$; (**e**) rolling motion $t = T/6$; (**f**) rolling case $t = T/4$; (**g**) rolling case $t = T/3$; (**h**) rolling case $t = T/2$.

When the subchannel reaches the equilibrium position ($t = 0$), the Euler force decreases to zero. The transverse velocity in the middle of the cross-section reaches the maximum, far more than that under static state, whereas the secondary vortices in the middle still prevail at the cross-section (Figure 5d). The maximum transverse velocity on the x'-axis occurs in the vicinity between Rod-2 and Rod-3 due to the promotion of boiling. Under the effect of boiling and rolling motion, transverse flow in the connected area comes up. However, the velocity profiles differ between the two sides of the x'-axis due to the combined effect of boiling and rolling motion. On the two sides of the y'-axis, the transverse velocity is slightly greater than that under static state, but rapidly decreases to zero near the middle of the secondary vortices.

Then, the subchannel rotates anticlockwise to the extreme position. The angular acceleration is positive during this process, and the Euler force on the fluid is negative along the x'-axis. Transverse velocity around the x'-axis decreases gradually and then increases reversely, and new secondary vortices emerge (Figure 5e). In the meantime, transverse velocity around the positive and negative sides of the y'-axis also decreases gradually, and the effect of boiling on the transverse flow field comes into view gradually.

When the subchannel reaches the extreme position ($t = T/4$), the Euler force increases to its maximum. The transverse flow field is similar to that under static state (Figure 5f). However, the transverse velocity at this time is far less than that at the other moments in the rolling motion process. The transverse velocity around the negative side of the x'-axis is strengthened due to the same direction of the transverse flow induced by boiling and rolling motion. In contrast, the transverse velocity around the positive side is weakened due to the opposite direction. Consequently, transverse velocity on the x'-axis decreases along the positive direction. The rolling motion almost does not affect the transverse flow around the positive and negative sides of the y'-axis. Two secondary vortices are formed in the rod gap, and the transverse velocity profile is the same as that under static state.

Next, the subchannel changes to rotate clockwise from the extreme position to the equilibrium position. During this process, the angular acceleration and the Euler force on the fluid still point to the x'-axis in the positive and negative directions, respectively. The transverse velocity in the middle of the cross-section increases, and two secondary vortices of reverse direction emerge (Figure 5g). When the subchannel reaches the equilibrium position ($t = T/2$) again, the Euler force is zero, and the transverse velocity in the middle of the cross-section increases to its maximum. The two transverse velocity profiles on the x'-axis are symmetrical around the axis of $x' = 0$ (Figure 5h).

The evolution of the transverse flow during the next half period is similar to those during the above-mentioned first half period.

3.1.2. Vapor Volume Fraction

Transverse flow under static state is very different from that under rolling motion, as is vapor volume fraction in the two-phase domain. Tendencies of vapor volume fraction in the domain are shown in Figure 6a, in which the vapor volume fraction is the area-averaged value in each domain of 0.10 m height. The results in four domains from CS-1.00 m to CS-1.40 m are displayed. At any moment, the vapor volume fraction increases along the flow direction in the static and the rolling cases. However, the tendencies of the vapor volume fraction are different. The vapor volume fraction in each domain is constant under static state, whereas it oscillates under rolling motion. The period of vapor volume fraction under rolling motion is almost half that of the rolling motion. Moreover, vapor volume fraction under rolling motion is always greater than that under static state for each domain.

Under static state, the vapor volume fraction in Quadrant-1 and Quadrant-2 are constant and approximately equal to each other (Figure 6b,d). The tendency of vapor volume fraction in each quadrant features a similar pattern to that of the whole domain. The vapor volume fraction increases along the flow direction in both quadrants. Under rolling motion, the tendency of the vapor volume fraction in each quadrant differs from that in the whole domain (Figure 6c,e). In each domain, vapor volume fraction in both quadrants oscillates with the period equal to that of rolling motion but is in phase opposition.

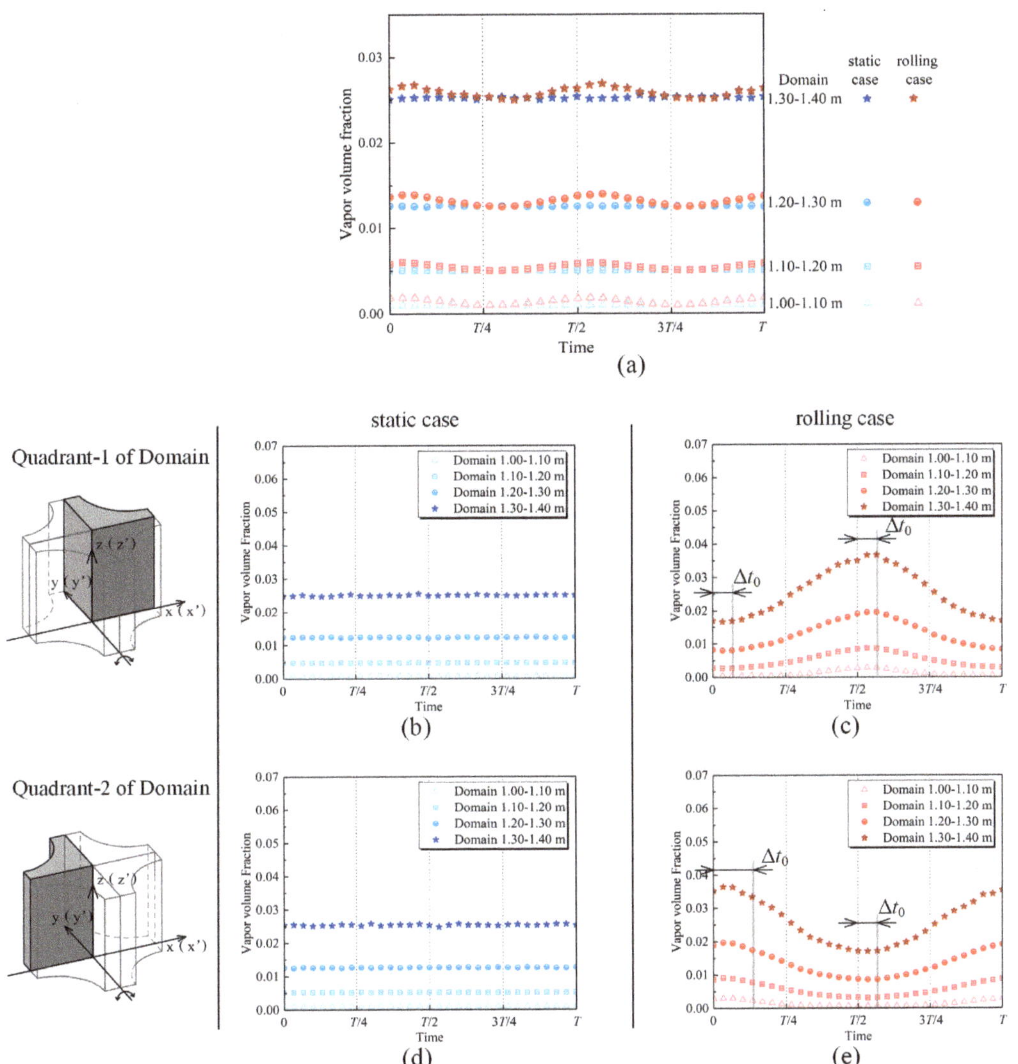

Figure 6. Instantaneous vapor volume fraction. (**a**) Domain; (**b**) Quadrant-1 in the static case; (**c**) Quadrant-1 in the rolling case; (**d**) Quadrant-2 in the static case; (**e**) Quadrant-2 in the rolling case.

The difference in vapor volume fraction between the static and rolling case is primarily attributed to the difference in transverse flow field. The transverse flow fields in both the quadrants are similar and constant with time under static state, whereas they differ significantly and oscillate periodically under rolling motion. The reason is explained below in chronological order.

The transverse flow pattern at the cross-section remains the same during the process when the subchannel rotates anticlockwise to the equilibrium position. Highly subcooled bulk fluid flows successively by Rod-1 and Rod-2, and the fluid temperature nearby Rod-1 is lower. As a result, the vapor volume fraction in Quadrant-1 is far less than that in Quadrant-

2 when the subchannel is in the equilibrium position ($t = 0$), when $t = \Delta t_0$. The vapor volume fraction in Quadrant-1 and Quadrant-2 reaches the minimum and the maximum, respectively. After that, the subchannel continues to rotate anticlockwise, and the transverse flow field begins to reverse. A portion of the bulk fluid and fluid in the connected area flow to the vicinity of Rod-2. Bubbles in Quadrant-2 condense quickly, leading to vapor volume fraction decreases therein. A portion of the bulk fluid still flows to the vicinity of Rod-1. However, the flow rate decreases with time, and the vapor volume fraction increases therein. When the subchannel reaches the extreme position ($t = T/4$), the vapor volume fractions in Quadrant-1 and Quadrant-2 are approximately equal to their values under static state. After that, the subchannel rotates clockwise, and the transverse flow pattern at the cross-section does not change after $t = T/3$ (Figure 5g,h). Highly subcooled bulk fluid flows successively by Rod-2 and Rod-1, and the bulk fluid temperature nearby Rod-1 is higher. As a result, the vapor volume fraction continues to decrease in Quadrant-1 and increase in Quadrant-2.

The tendencies of vapor volume fraction in both quadrants in the next half period reversed those in the above-mentioned first half period.

Noteworthy, in the rolling case, the vapor volume fraction in Quadrant-1 and Quadrant-2 oscillates around that under static state and is in phase opposition. The summation of the vapor volume fraction in each quadrant, i.e., the vapor volume fraction in a domain, is equal to or greater than that in the static case. This is because the increment under rolling motion is larger than the decrement all the time. Especially in Domain 1.0–1.1 m, the vapor volume fraction in each quadrant is equal to or greater than that under static state due to the lower position where the boil begins.

3.1.3. Vapor Adhesion Ratio

For subcooled flow boiling in the subchannel, the vapor volume fraction remains small in the computation domain and typically no more than a few percent. However, the bubbles adjacent to the wall may crowd sufficiently, and a growing bubble layer may form eventually. The departure from nucleate boiling (DNB) occurs without a sustained macroscopic contact between the liquid and the surface, even though the bulk flow in the subchannel may still be highly subcooled. Predictions of the vapor volume fraction and vapor adhesion are essential. The vapor adhesion ratio at the wall can be defined as

$$R = \frac{\int_{\text{vapor}} dA}{\int_{\text{wall}} dA} \tag{20}$$

The denominator denotes the area of the wall, and the numerator denotes the area of the wall covered with vapor.

The tendencies of the vapor adhesion ratio at the wall are shown in Figure 7a. The vapor adhesion ratio is an area-averaged value at the wall, including both the heated and unheated surface in each domain 0.1 m high. It is defined in this way because the bubbles can not only be formed adjacent to the heated surface but also be pushed to the unheated surface. The vapor adhesion ratios increase along the flow direction under both static and rolling cases. The vapor adhesion ratio of Segment 1.00–1.10 m under rolling motion is greater than that under static state, whereas the vapor adhesion ratio of Segment 1.10–1.40 m is lower than that under static state.

For the static case, all the walls have axial symmetry, and the vapor adhesion ratio at them is identical and stable (Figure 7b,d). For the rolling case, the vapor adhesion ratios at both walls oscillate with a period equal to that of rolling motion (Figure 7c,e). Interestingly, the vapor adhesion ratio at Wall-2 is essentially constant when the vapor adhesion ratio at Wall-1 decreases or increases, and vice versa.

The vapor adhesion ratio at each wall increases along the flow direction. The effect of the transverse flow on the vapor adhesion ratio is slight at the lower position (Segment 1.00–1.20 m) under rolling motion. In Segment 1.00–1.10 m, the variation in the vapor adhesion ratio with time can be divided into three phases on both walls: increase, constant, and decrease. Furthermore, the vapor adhesion ratios under rolling motion are always greater than those under static state on both walls. They oscillate with the period half as much as that of rolling motion. In Segment 1.10–1.20 m, the vapor adhesion ratios under rolling motion on both walls oscillate around that under static state.

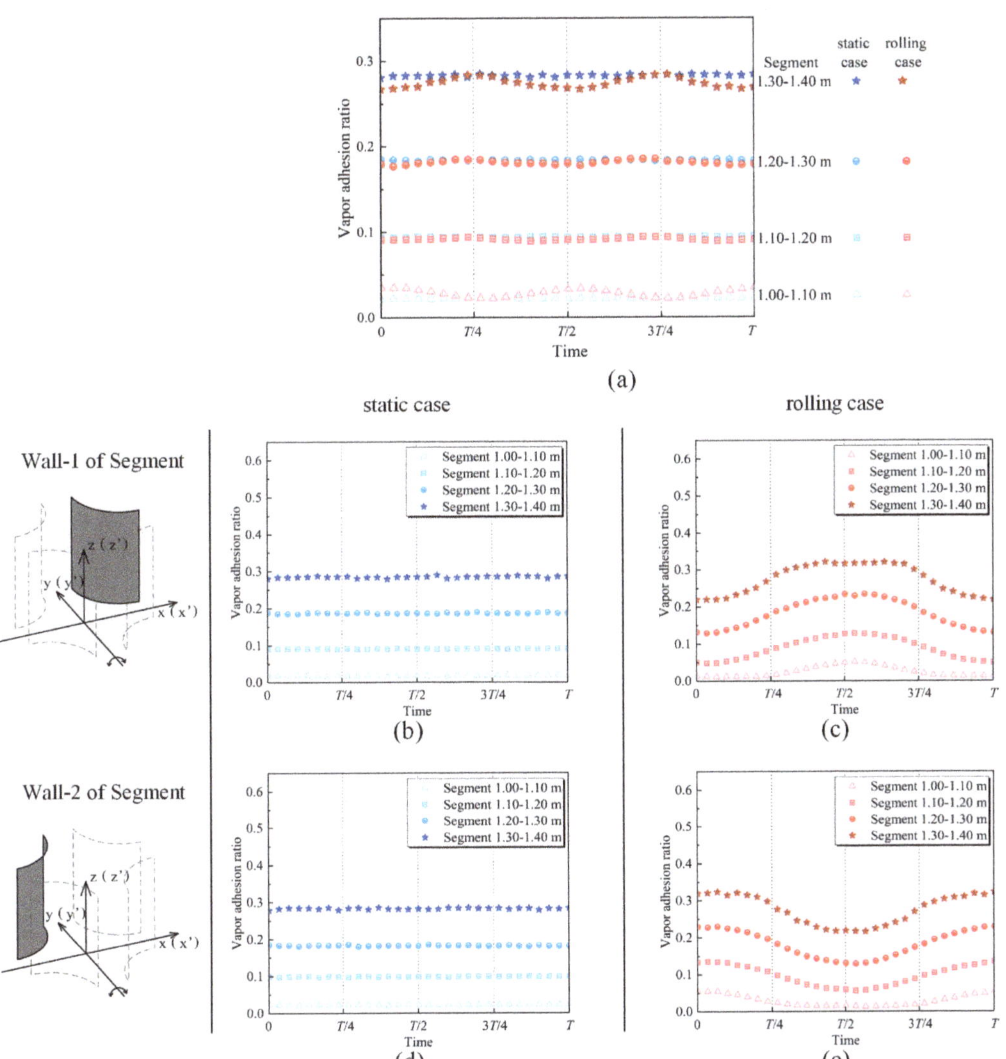

Figure 7. Instantaneous vapor adhesion ratio. (**a**) Segment; (**b**) Wall-1 in the static case; (**c**) Wall-1 in the rolling case; (**d**) Wall-2 in the static case; (**e**) Wall-2 in the rolling case.

The tendencies of the vapor adhesion ratio are different at the upper position (Segment 1.20–1.40 m) due to the combined effects of the vapor volume fraction and the transverse flow. At the beginning of rolling motion ($t = 0$), the vapor adhesion ratio at Wall-1 is less than that under static state due to transverse flow. Then, it gradually increases to approximately the value under static state when the subchannel reaches the extreme position ($t = T/4$). After that, the incoming fluid temperature in the vicinity of Rod-1 begins to increase, and more fluid reaches saturated to boil. In the meantime, the vapor volume fraction in Quadrant-1 and the vapor adhesion ratio at Wall-1 increase.

When the subchannel rotates anticlockwise from the extreme to the equilibrium position, the vapor volume fraction in Quadrant-1 increases continuously, and bubbles on the heated surface shift along the transverse flow. Finally, a great quantity of bubbles accumulate near the boundary between the heated surface and the unheated surface by the entrainment and flush of transverse flow. The vapor layer thickness increases as the vapor volume fraction increases, while the vapor adhesion ratio is constant (Figure 7d). The subcooled fluid in the middle of the cross-section and the gap between Rod-1 and Rod-2 flows to Rod-1 after the subchannel reaches the equilibrium position ($t = T/2$). The vapor volume fraction in Quadrant-1 decreases gradually. The bubbles, far from the heated surface in the boundary between the heated surface and the unheated surface, are apt to be affected by transverse flow and begin to condense. Therefore, the maximum vapor layer thickness at Wall-1 decreases while the vapor adhesion ratio is constant.

When the subchannel reaches the extreme position again ($t = 3T/4$), the transverse velocity in the vicinity of Rod-1 begins to increase. After that, the vapor volume fraction in Quadrant-1 decreases gradually. The bubbles shift to the center of the heated surface slowly, which leads to the decrease in the vapor adhesion ratio at Wall-1.

The tendencies of the vapor adhesion ratio at other walls follow the same pattern as that at Wall-1. In conclusion, although the vapor volume fraction under rolling motion is always higher than that under static state at the upper position of the flow domain, the transverse flow tends to cause bubble accumulation, and the vapor adhesion ratio is less than that under static state all the time.

3.1.4. Phase Distribution

For PWRs at normal operation, local boiling occurs at a higher elevation of the reactor core. Vapor bubbles grow attached to the fuel rod while the bulk flow is still highly subcooled. As the height increases, multiple bubbles coalesce to form thin and discontinuous vapor layers. Therefore, the flow pattern in the present study is quite different from that in saturated boiling. In addition, the phase distribution in the subchannel is significantly different from that in the circular and rectangular tubes under the effect of the geometrical configuration.

The vapor–liquid phase distribution in Quadrant-1 is shown in Figure 8. Only the vapor phase is shown in the figure, with color indicating the perpendicular distance between the interface and the heated surface. The region inside the black dashed line is the heated surface. The regions between the continuous and dashed lines are the rectangular unheated surfaces adjacent to the rod.

Flow and heat transfer in the subchannel are stable under static state. The OSV occurs around CS-1.05 m. Even though the heat flux of the fuel rod is uniform, fluid along both sides of the heated surface reaches saturated earlier under the effect of the channel sectional configuration (Figure 8a).

As the height increases, fluid along both sides and the center of the heated surface begins to boil (Figure 8b). The bubble is not the typical sphere but is long and narrow due to the transverse flow (Figure 5a). The fluid near the heated surface flows from both sides to the center, and the bubbles will be elongated to incline towards the center of the heated surface. Some bubbles will break into multiple small bubbles to merge with bubbles in the center of the heated surface. Bubbles in the center of the heated surface will grow continuously and bridge bubbles on both sides.

Further downstream, two vapor columns come into being due to more bubbles accumulating on both sides of the heated surface (Figure 8c). It should be emphasized that part of the vapor column herein is separated from the heated surface by liquid and does not cover the heated surface absolutely (Figure 8d). Vapor near the surface is submerged in the boundary layer with low velocity, and vapor far away from the surface flows downstream with high velocity. They detach from each other due to the velocity difference. Vapor far away from the surface interconnects together as a column, whereas the vapor phase and the liquid phase are separated near the surface.

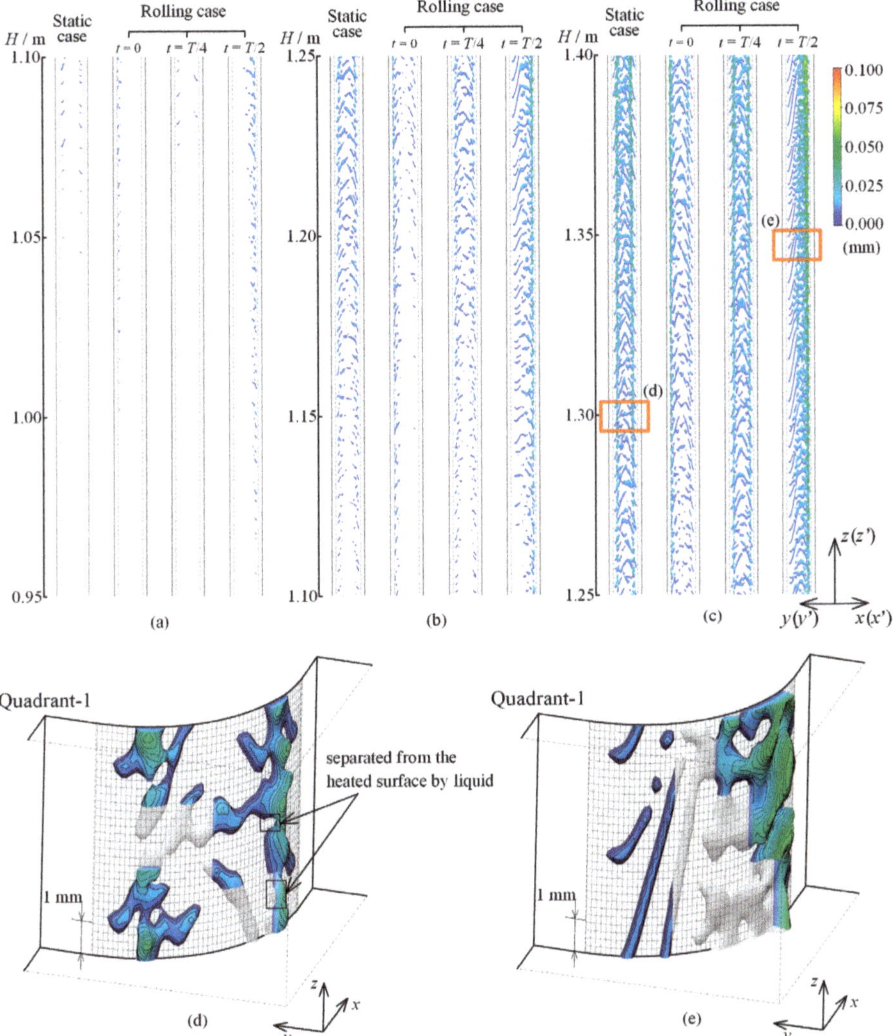

Figure 8. Vapor–liquid phase distribution in Quadrant-1: (**a**) 0.95–1.10 m; (**b**) 1.10–1.25 m; (**c**) 1.25–1.40 m. (**d**) A detail of the static case; (**e**) a detail of the rolling case.

Under rolling motion, the vapor–liquid phase distribution in the subchannel varies with time due to transverse flow, which features the following characteristics. When the subchannel reaches the equilibrium position ($t = 0$ or $t = T/2$), which is the same as that under static state, the phase distribution is significantly different from that under static state. The OSV point under rolling motion is much lower than that under static state (Figure 8a). Further downstream, vapor bubbles are swept by the transverse flow and the amount of the vapor increases along the direction of the transverse flow. Although the transverse flow fields are similar when $t = 0$ and $t = T/2$, the phase distributions are slightly different. Different directions of the transverse flow not only result in different positions where bubbles accumulate but also affect the vapor volume fraction. When $t = T/2$, transverse flow heated by Rod-2 then flows to Rod-1, causing more liquid to boil (Figure 8e). Bubbles on the left side of the heated surface grow in length and eventually break into discrete bubbles. Multiple bubbles on the right side coalesce and collect at the boundary of the heated and unheated surface. When the subchannel reaches the extreme position ($t = T/4$ or $t = 3T/4$), the vapor–liquid phase distribution is similar to that under static state due to a similar transverse flow field.

3.2. Sensitivity Analysis

3.2.1. Transverse Flow

The effects of the amplitude on the transverse flow are shown in Figure 9, taking the transverse velocity in CS-1.40 m as an example. The transverse flow velocity distributions are similar for different amplitudes. The rolling motion dominates the transverse flow in the middle of the cross-section at any time. Therefore, the Euler force increases as the amplitude increases. So, the transverse velocity in the middle of the cross-section increases. The transverse flow in other locations is dominated by both boiling and rolling motion.

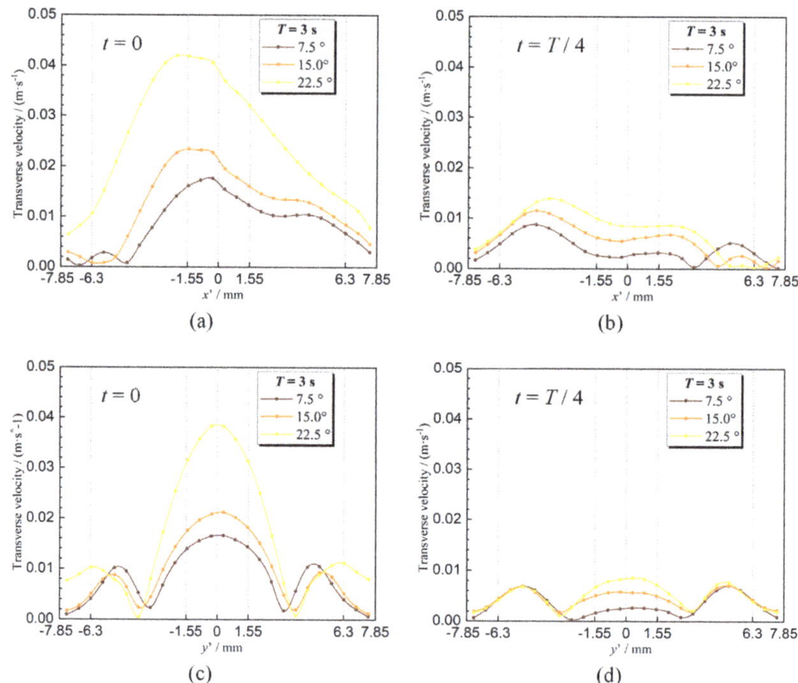

Figure 9. Effect of the amplitude on the transverse velocity. Transverse velocity on the x'-axis at (**a**) $t = 0$ and (**b**) $t = T/4$; (**c**) transverse velocity on the y' -axis at (**c**) $t = 0$ and (**d**) $t = T/4$.

The transverse velocity in the middle of the cross-section is in the positive direction of the x'-axis when the subchannel is in the equilibrium position ($t = 0$). On the negative side of the x'-axis, the transverse flow is in the negative x'-axis direction. For the small amplitude (7.5°), the transverse velocity induced by rolling motion is small, and the effect of boiling on the transverse flow is prominent. At this time, the transverse velocity profile is similar to that under static state. For the large amplitude (22.5°), the transverse velocity induced by rolling motion is large, and the effect of the rolling motion is dominant. The closer to the middle of the cross-section, the greater the transverse velocity. Figure 10 compares the transverse flow with different amplitudes of rolling motion. The secondary vortices induced by boiling still exist around the negative x'-axis when the amplitudes are 7.5° and 15.0° but vanish thoroughly at 22.5° (the red box in Figure 10).

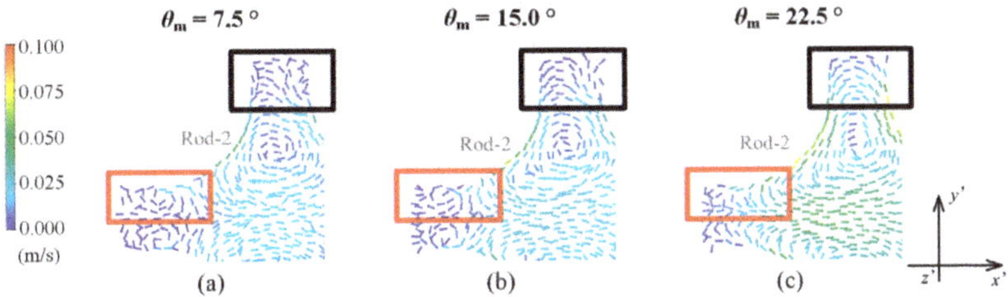

Figure 10. Transverse flow field with different amplitudes ($t = 0$). (**a**) $\theta_m = 7.5°$; (**b**) $\theta_m = 15.0°$; (**c**) $\theta_m = 22.5°$.

On the positive x'-axis, the transverse flow induced by boiling is superimposed on that induced by rolling motion due to identical direction. The transverse velocity here increases as the amplitude increases.

On the y'-axis, the transverse flow on the positive side is identical to that on the negative side. If the amplitude of rolling motion is small (7.5°), the transverse velocity induced by rolling motion is small, and the transverse flow on the y'-axis is dominated by both boiling and rolling motion. The velocity profiles are approximately the same, but the magnitude is higher than that under static state. If the amplitude of rolling motion is large (22.5°), the effect of rolling motion on the transverse flow is prominent. The transverse flow induced by rolling is higher, especially on both ends of the y'-axis. As can be seen from the black box in Figure 10, the secondary vortex is intensified gradually, and the center approaches the y'-axis as the amplitude increases. When the amplitude is 22.5°, the location where the transverse velocity on y′-axial equals zero is the center of the secondary vortex.

The transverse velocity in the middle of the cross-section is in the negative direction of the x'-axis when the subchannel is in the extreme position ($t = T/4$). On the x'-axis, the transverse flow is dominated by both boiling and rolling motion. The transverse flow induced by boiling remains the same throughout. Hence, the rolling motion enhances the transverse flow on the negative x'-axis and mitigates it on the positive side. As the amplitude increases, the transverse velocities on the negative and positive sides of the x'-axis are in reverse. On the y'-axis, the transverse flow is only dominated by boiling and independent of the amplitude of rolling motion.

The effects of the period on the transverse flow are depicted in Figure 11. The transverse velocity profiles are the same for different periods. The transverse flow in the middle of the cross-section is dominated by rolling motion all the time. Therefore, as the period decreases, the Euler force increases, as does the transverse velocity in the middle of the cross-section. In the other location of the cross-section, the transverse flow is dominated by both the boiling and rolling motion.

When the subchannel is in the equilibrium position ($t = 0$), the transverse flows on both the x'-axis and y'-axis are dominated by the rolling motion. Therefore, the transverse flow velocity in the whole cross-section decreases as the period decreases, except that in the center of the secondary vortex is zero.

When the subchannel is in the extreme position ($t = T/4$), the effect of boiling on the transverse flow comes into view. On the x'-axis, the transverse velocity on the negative side increases while that on the positive side decreases as the period decreases. The transverse flow on the y'-axis is only dominated by boiling and independent of the period.

In summary, the Euler force is the fundamental reason for the variation in the transverse flow under rolling motion. It depends on the amplitude and period of rolling motion. The Euler force increases with the amplitude, decreases with the period, and vice versa. Consequently, the tendency of the transverse flow is the same when the amplitude increases or the period decreases.

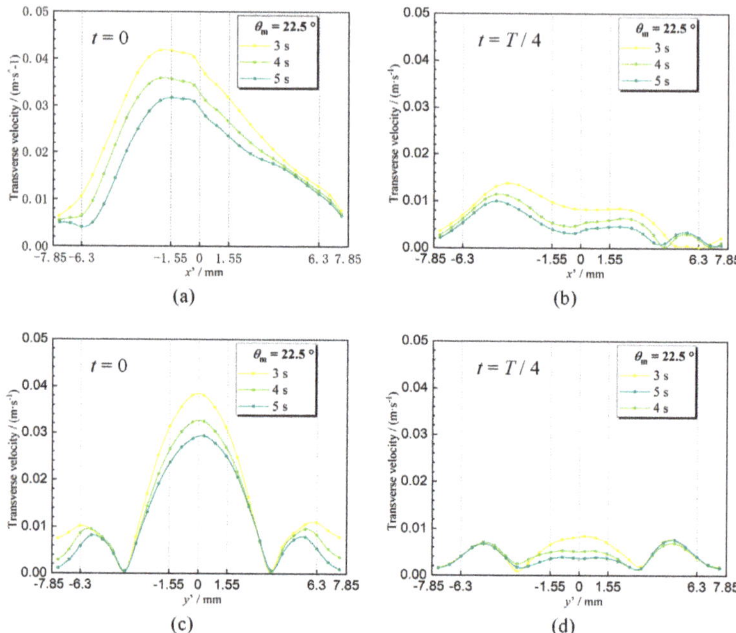

Figure 11. Effect of the period on the transverse velocity. Transverse velocity on the x'-axis at (**a**) $t = 0$ and (**b**) $t = T/4$; (**c**) transverse velocity on the y'-axis at (**c**) $t = 0$ and (**d**) $t = T/4$.

3.2.2. Vapor Volume Fraction

The tendencies of vapor volume fraction are similar for different amplitudes of rolling motion in each domain (Figure 12a). The vapor volume fraction in each domain approaches the maximum when the subchannel deviates from the equilibrium position (about $t = 0$ or $t = T/2$), and the minimum when it deviates from the extreme position (about $t = T/4$ or $t = 3T/4$). The period of the vapor volume fraction is half that of rolling motion. As the amplitude of rolling motion increases, the maximum of the vapor volume fraction in each domain increases. For Domain 1.30–1.40 m, the maximum of the vapor volume fraction under rolling motion is 2.5%, 7.3%, and 17.5% larger than the time-averaged vapor volume fraction under static state when the amplitude is 7.5°, 15.0°, and 22.5°, respectively. However, the minimum is constantly equivalent to that in the same domain under static state.

The vapor volume fraction in different quadrants is analyzed, taking the vapor volume fraction in Domain 1.30–1.40 m as an example (Figure 12b,c). The tendencies of the vapor volume fraction in Quadrant-1 are similar for different amplitudes of rolling motion. The tendencies of the vapor volume fraction in Quadrant-2 are the same as those in Quadrant-1 but different in phase by 180°. When the subchannel deviates from the equilibrium position (about $t = 0$ or $t = T/2$), the difference between the vapor volume fraction in Quadrant-1 and Quadrant-2 increases as the amplitude of rolling motion increases. When the subchannel deviates from the extreme position (about $t = T/4$ or $t = 3T/4$), the vapor volume fraction in Quadrant-1 or Quadrant-2 is independent of the amplitude of the rolling motion. Therefore, the maximum of the vapor volume fraction in each domain increases while the minimum is unchanged as the amplitude of rolling motion increases.

Figure 13 depicts the vapor volume fraction variation with the rolling motion period. As discussed previously, the variation in the vapor volume fraction fundamentally results from the variation in the transverse flow induced by the Euler force. When the amplitude of rolling motion increases or the period decreases, the variation in the transverse flow is identical, and so is the vapor volume fraction.

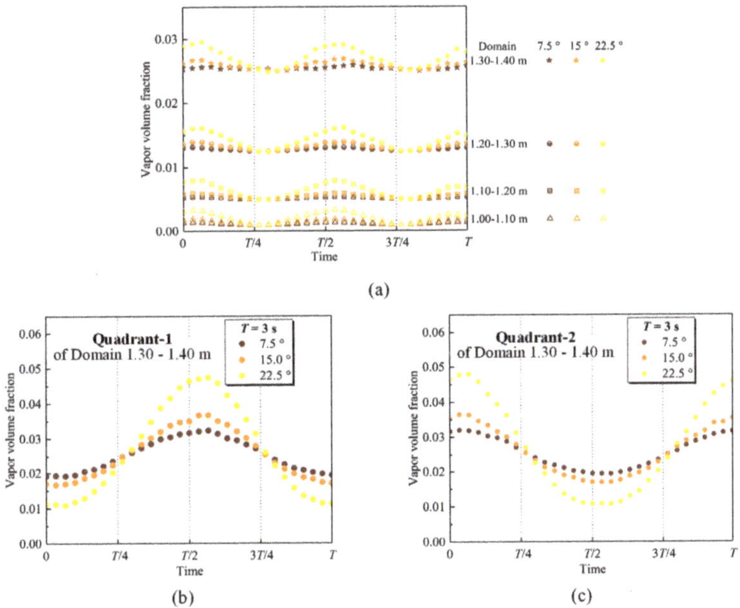

Figure 12. Effect of the amplitude on the vapor volume fraction. (**a**) Domain; (**b**) Quadrant-1 of Domain 1.30–1.40 m; (**c**) Quadrant-2 of Domain 1.30–1.40 m.

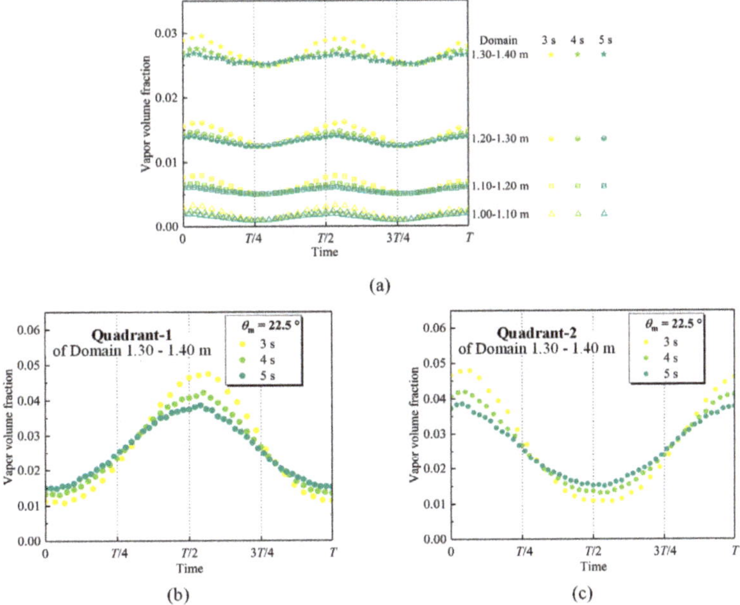

Figure 13. Effect of the period on the vapor volume fraction. (**a**) Domain; (**b**) Quadrant-1 of Domain 1.30–1.40 m; (**c**) Quadrant-2 of Domain 1.30–1.40 m.

3.2.3. Vapor Adhesion Ratio

The tendencies of the vapor adhesion ratio are similar for different amplitudes of rolling motion at each segment (Figure 14a). The vapor adhesion ratio at each segment approaches the minimum when the subchannel deviates from the equilibrium position (about $t = 0$ or $t = T/2$), and the maximum when it deviates from the extreme position (about $t = T/4$ or $t = 3T/4$). The period of the vapor adhesion ratio is half that of rolling motion. As the amplitude of rolling motion increases, the minimum of the vapor adhesion ratio at each segment decreases. For Segment 1.30–1.40 m, the minimum of the vapor adhesion ratio under rolling motion is 18.0%, 23.0%, and 45.9% smaller than the time-averaged vapor adhesion ratio under static state when the amplitude is 7.5°, 15.0°, and 22.5°, respectively. Nevertheless, the maximum is equivalent to that in the same segment under static state.

The vapor adhesion ratio at different walls is analyzed, taking the vapor adhesion ratio at Segment 1.30–1.40 m as an example (Figure 14b,c). The tendencies of the vapor adhesion ratio at Wall-1 are similar for different amplitudes of rolling motion. The minimum of the vapor adhesion ratio decreases, and the maximum is constant as the amplitude of rolling motion increases. The tendencies of the vapor adhesion ratio at Wall-2 and Wall-1 are the same but different by a relative 180° phase shift. When the subchannel deviates from the equilibrium position (about $t = 0$ or $t = T/2$), the difference between the vapor adhesion ratio at Wall-1 and Wall-2 increases as the amplitude of rolling motion increases. When the subchannel deviates from the extreme position (about $t = T/4$ or $t = 3T/4$), the vapor adhesion ratio at Wall-1 or Wall-2 is independent of the amplitude of the rolling motion. Therefore, the minimum of the vapor adhesion ratio in each segment decreases while the maximum is unchanged as the amplitude of rolling motion increases.

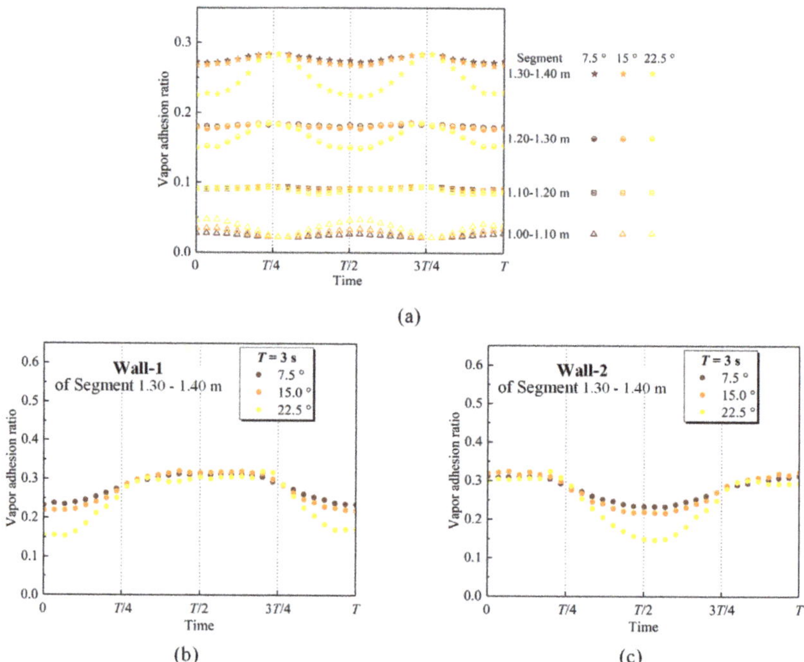

Figure 14. Effect of the amplitude on the vapor adhesion ratio. (**a**) Segment; (**b**) Wall-1 of Segment 1.30–1.40 m; (**c**) Wall-2 of Segment 1.30–1.40 m.

Figure 15 depicts the vapor adhesion ratio variation with the rolling motion period. In addition to the vapor volume fraction, the variation in the transverse flow induced by the Euler force fundamentally results in the variation in the vapor adhesion ratio. Therefore, when the amplitude of rolling motion increases or the period decreases, the variation in the transverse flow is identical, as is the vapor adhesion ratio. Figure 16 shows the effect of the amplitude and the period on the vapor–liquid phase distribution in Quadrant-1 ($t = T/2$). The bubbles accumulate on the boundary of the heated and unheated surface when the subchannel is in the equilibrium position. When the amplitude increases or the period decreases, the bubble thickness increases, but the vapor adhesion ratio is basically constant.

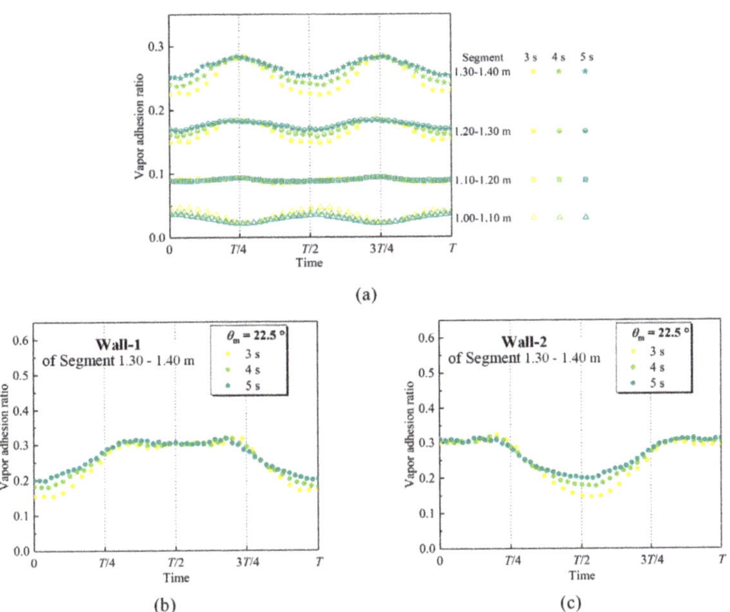

Figure 15. Effect of the period on the vapor adhesion ratio. (**a**) Segment; (**b**) Wall-1 of Segment 1.30–1.40 m; (**c**) Wall-2 of Segment 1.30–1.40 m.

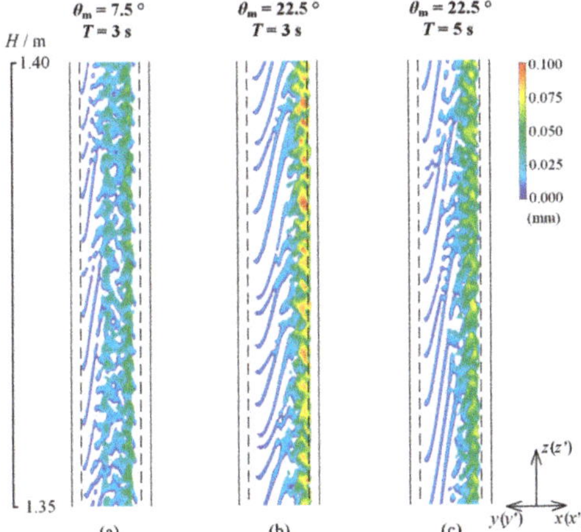

Figure 16. Phase distribution with different amplitudes and periods ($t = T/2$). (**a**) $\theta_m = 7.5°$, $T = 3$ s; (**b**) $\theta_m = 22.5°$, $T = 3$ s; (**c**) $\theta_m = 22.5°$, $T = 5$ s.

4. Discussion

In the present study, we investigated the subcooled flow boiling in the subchannel under rolling motion following the trigonometric function. The tendency of physical parameters, including the transverse flow, volume fraction, and vapor adhesion ratio, was explored along the flow direction. The phase distribution in the two-phase flow domain is presented in this paper. The effects and mechanisms of the rolling motion parameters on the two-phase flow in the subchannel are analyzed. The main conclusions are as follows.

(1) The transverse flow under rolling motion is a composition of the boiling-induced and motion-induced transverse flow. The transverse flow induced by rolling motion fundamentally results from the Euler force in the non-inertial reference frame. The transverse flow in the center of the cross-section increases with the amplitude of the rolling motion and decreases with the period. The transverse flow in the other position of the cross-section depends on the rolling motion parameters and the position of the subchannel.

(2) The location of the OSV point and the vapor volume fraction in the two-phase flow domain varies with time because of the variation in the transverse flow. When the subchannel is around the equilibrium position, the volume fraction reaches the maximum, which increases with the amplitude of the rolling motion and decreases with the period. When the subchannel is about the extreme position, the volume fraction reaches the minimum, which is insensitive to the rolling motion parameters.

(3) In the present study, the vapor adhesion ratio upstream of CS-1.20 m is only related to the volume fraction, while the downstream is affected by both the volume fraction and the transverse flow. For the flow domain downstream of CS-1.20 m, the vapor adhesion ratio in the downstream reaches the minimum when the subchannel is around the equilibrium position. This minimum decreases with the amplitude of the rolling motion and increases with the period. The vapor adhesion ratio in the downstream reaches the maximum when the subchannel is about the extreme position. The maximum is insensitive to the rolling motion parameters.

(4) When the rolling motion parameters change, the phase distribution is similar at the same time under rolling motion. In the present study, bubbles slide to the boundary of the heated and unheated surface. The bubble thickness at this boundary increases with the amplitude of the rolling motion and decreases with the period.

Author Contributions: Conceptualization, Y.L., X.R. and N.W.; Methodology, Y.L. and X.Y.; Software, X.C., Z.N. and X.Y.; Validation, Y.L. and Z.N.; Formal Analysis, Y.L. and X.C.; Investigation, Y.L., Z.N. and X.Y.; Resources, X.R. and N.W.; Writing—Original Draft Preparation, Y.L., X.C., X.R. and N.W.; Writing—Review and Editing, Y.L., X.C., X.R. and N.W.; Supervision, N.W.; Project Administration, N.W. All authors have read and agreed to the published version of the manuscript.

Funding: This research received no external funding.

Institutional Review Board Statement: Not applicable.

Informed Consent Statement: Not applicable.

Data Availability Statement: Not applicable.

Acknowledgments: The scientific calculations in this paper were performed on the HPC Cloud Platform of Shandong University.

Conflicts of Interest: The authors declare no conflict of interest.

Nomenclature

General Symbols

A	area, m^2
a	acceleration, m/s^2
a_τ	Euler acceleration, m/s^2
a_n	centrifugal acceleration, m/s^2
a_c	Coriolis acceleration, m/s^2

E	energy
F	force, N
F_τ	Euler force, N
F_n	centrifugal force, N
F_c	Coriolis force, N
F_s	surface tension, N/m
g	gravity, m/s^2
h_{lv}	latent heat, J/kg
H	height, m
k	conductive coefficient, W/(m· K)
v	velocity, m/s
m	mass, kg
$\dot{m}$	mass transfer rate, $kg/(m^3 \cdot s)$
$\vec{n}$	boundary normal vector
p	pressure, Pa
$\vec{r}$	position vector
R	vapor adhesion ratio
S	inertial frame of reference
S'	non-inertial frame of reference
S_m	mass source term, $kg/(m^3 \cdot s)$
S_h	energy source term, W/m^3
T	period of the rolling motion, s
T''	temperature, K
t	time, s
Greek Letters	
α	volume fraction
β	(evaporation and condensation) coefficient, 1/s
θ	angle, rad
θ_m	amplitude of the rolling motion, rad
μ	viscosity, Pa·s
ρ	density, kg/m^3
ω	angular velocity, rad/s
ε	angular acceleration, rad/s^2
Superscript	
$'$	in the non-inertial reference frame
Subscripts	
l	liquid
sat	saturation state
v	vapor
x, y, and z	x-, y-, and z-axes

References

1. Yan, C.; Yan, C.; Sun, L.; Xing, D.; Wang, Y. Investigation of the interfacial parameter distribution in a bubbly flow in a narrow rectangular channel under inclined and rolling conditions. *Prog. Nucl. Energy* **2014**, *73*, 64–74. [CrossRef]
2. Tanjung, E.F.; Kim, B.J.; Jo, D. Effects on pool boiling critical heat flux (CHF) with different direction and magnitude of additional accelerations due to rolling. *Ann. Nucl. Energy* **2021**, *154*, 108095. [CrossRef]
3. Hwang, J.-S.; Lee, Y.-G.; Park, G.-C. Characteristics of critical heat flux under rolling condition for flow boiling in vertical tube. *Nucl. Eng. Des.* **2012**, *252*, 153–162. [CrossRef]
4. Li, S.; Tan, S.; Gao, P.; Xu, C. Experimental research of bubble number density and bubble size in narrow rectangular channel under rolling motion. *Nucl. Eng. Des.* **2014**, *268*, 41–50. [CrossRef]
5. Li, S.; Tan, S.; Xu, C.; Gao, P. Visualization study of bubble behavior in a subcooled flow boiling channel under rolling motion. *Ann. Nucl. Energy* **2015**, *76*, 390–400. [CrossRef]
6. Kharangate, C.R.; Mudawar, I. Review of computational studies on boiling and condensation. *Int. J. Heat Mass Transf.* **2017**, *108*, 1164–1196. [CrossRef]
7. Choudhury, R.; Das, U.J.; Ceruti, A.; Piancastelli, L.; Frizziero, L.; Zanuccoli, G.; Daidzic, N.; Rocchi, I.; Casano, G.; Piva, S. Visco-elastic effects on the three dimensional hydrodynamic flow past a vertical porous plate. *Int. Inf. Eng. Technol. Assoc.* **2013**, *31*, 1–8. [CrossRef]

8. Valizadeh Yaghmourali, Y.; Ahmadi, N.; Abbaspour-sani, E. A thermal-calorimetric gas flow meter with improved isolating feature. *Microsyst. Technol.* **2016**, *23*, 1927–1936. [CrossRef]
9. Thome, J.R. (Ed.) *Encyclopedia of Two-Phase Heat Transfer and Flow I Fundamentals and Methods Volume 4: Special Topics in Pool and Flow Boiling*; World Scientific Publishing: Singapore, 2015; p. 176.
10. Jia, H.W.; Zhang, P.; Fu, X.; Jiang, S.C. A numerical investigation of nucleate boiling at a constant surface temperature. *Appl. Therm. Eng.* **2015**, *88*, 248–257. [CrossRef]
11. Cheng, N.; Guo, Y.; Peng, C. A numerical simulation of single bubble growth in subcooled boiling water. *Ann. Nucl. Energy* **2019**, *124*, 179–186. [CrossRef]
12. Bahreini, M.; Ramiar, A.; Ranjbar, A.A. Numerical simulation of bubble behavior in subcooled flow boiling under velocity and temperature gradient. *Nucl. Eng. Des.* **2015**, *293*, 238–248. [CrossRef]
13. Jeon, S.-S.; Kim, S.-J.; Park, G.-C. Numerical study of condensing bubble in subcooled boiling flow using volume of fluid model. *Chem. Eng. Sci.* **2011**, *66*, 5899–5909. [CrossRef]
14. Guo, L.; Zhang, S.; Hu, J. Flow boiling heat transfer characteristics of two-phase flow in microchannels. *AIP Adv.* **2022**, *12*, 055219. [CrossRef]
15. Cheng, N.; Yu, S.; Xiao, J.; Peng, C. Numerical simulation of single bubble growth in vertical rectangular narrow flow channels. *Nucl. Eng. Des.* **2022**, *391*, 111749. [CrossRef]
16. Pattamatta, A.; Freystein, M.; Stephan, P. A parametric study on phase change heat transfer due to Taylor-Bubble coalescence in a square minichannel. *Int. J. Heat Mass Transf.* **2014**, *76*, 16–32. [CrossRef]
17. Liu, Z.; Sunden, B.; Wu, H. Numerical Modeling of Multiple Bubbles Condensation in Subcooled Flow Boiling. *J. Therm. Sci. Eng. Appl.* **2015**, *7*, 031003. [CrossRef]
18. Lee, J.; O'Neill, L.E.; Mudawar, I. 3-D computational investigation and experimental validation of effect of shear-lift on two-phase flow and heat transfer characteristics of highly subcooled flow boiling in vertical upflow. *Int. J. Heat Mass Transf.* **2020**, *150*, 119291. [CrossRef]
19. Kuang, Y.W.; Wang, W.; Zhuan, R.; Yi, C.C. Simulation of boiling flow in evaporator of separate type heat pipe with low heat flux. *Ann. Nucl. Energy* **2015**, *75*, 158–167. [CrossRef]
20. Yin, X.; Tian, Y.; Zhou, D.; Wang, N. Numerical study of flow boiling in an intermediate-scale vertical tube under low heat flux. *Appl. Therm. Eng.* **2019**, *153*, 739–747. [CrossRef]
21. Devahdhanush, V.S.; Lei, Y.; Chen, Z.; Mudawar, I. Assessing advantages and disadvantages of macro- and micro-channel flow boiling for high-heat-flux thermal management using computational and theoretical/empirical methods. *Int. J. Heat Mass Transf.* **2021**, *169*, 120787. [CrossRef]
22. Darshan, M.B.; Kumar, R.; Das, A.K. Numerical study of interfacial dynamics in flow boiling of R134a inside smooth and structured tubes. *Int. J. Heat Mass Transf.* **2022**, *188*, 122592. [CrossRef]
23. Zhuan, R.; Wang, W. Flow pattern of boiling in micro-channel by numerical simulation. *Int. J. Heat Mass Transf.* **2012**, *55*, 1741–1753. [CrossRef]
24. Yang, Z.; Peng, X.F.; Ye, P. Numerical and experimental investigation of two phase flow during boiling in a coiled tube. *Int. J. Heat Mass Transf.* **2008**, *51*, 1003–1016. [CrossRef]
25. Amidu, M.A.; Addad, Y.; Riahi, M.K. A hybrid multiphase flow model for the prediction of both low and high void fraction nucleate boiling regimes. *Appl. Therm. Eng.* **2020**, *178*, 115625. [CrossRef]
26. Bahreini, M.; Ramiar, A.; Ranjbar, A.A. Numerical simulation of subcooled flow boiling under conjugate heat transfer and microgravity condition in a vertical mini channel. *Appl. Therm. Eng.* **2017**, *113*, 170–185. [CrossRef]
27. Lee, J.; Mudawar, I.; Hasan, M.M.; Nahra, H.K.; Mackey, J.R. Experimental and computational investigation of flow boiling in microgravity. *Int. J. Heat Mass Transf.* **2022**, *183*, 122237. [CrossRef]
28. Wang, L.; Wang, J.; Li, Y. CFD study on film boiling features of cryogenic fluid influenced by heat structures and gravity levels. *Cryogenics* **2022**, *124*, 103455. [CrossRef]
29. Wei, J.-H.; Pan, L.-M.; Chen, D.-Q.; Zhang, H.; Xu, J.-J.; Huang, Y.-P. Numerical simulation of bubble behaviors in subcooled flow boiling under swing motion. *Nucl. Eng. Des.* **2011**, *241*, 2898–2908. [CrossRef]
30. Brackbill, J.U.; Kothe, D.B.; Zemach, C. A continuum method for modeling surface tension. *J. Comput. Phys.* **1992**, *100*, 335–354. [CrossRef]
31. Lemmon, E.W.; Bell, I.H.; Huber, M.L.; McLinden, M.O. *"Thermophysical Properties of Fluid Systems" in NIST Chemistry WebBook*; NIST Standard Reference Database Number 69; Linstrom, P.J., Mallard, W.G., Eds.; National Institute of Standards and Technology: Gaithersburg, MD, USA, 2022; p. 20899. Available online: https://webbook.nist.gov/chemistry/fluid/ (accessed on 15 January 2022).
32. Rubin, A.; Schoedel, A.; Avramova, M.; Utsuno, H.; Bajorek, S.; Velazquez-Lozada, A. *OECD/NRC Benchmark Based on NUPEC PWR Sub-Channel and Bundle Test (PSBT). Volume I: Experimental Database and Final Problem Specifications*; Nuclear Energy Agency of the OECD (NEA): Paris, France, 2012.
33. Prandtl, L. *Essentials of Fluid Dynamics: With Applications to Hydraulics Aeronautics, Meteorology, and Other Subjects*; Hafner Publishing Company: New York, NY, USA, 1952.
34. Bieder, U.; Falk, F.; Fauchet, G. LES analysis of the flow in a simplified PWR assembly with mixing grid. *Prog. Nucl. Energy* **2014**, *75*, 15–24. [CrossRef]

Article

Parametric Study on Thermo-Hydraulic Performance of NACA Airfoil Fin PCHEs Channels

Wei Wang [1,*], Liang Ding [1], Fangming Han [2], Yong Shuai [1], Bingxi Li [1] and Bengt Sunden [3,*]

[1] School of Energy Science and Engineering, Harbin Institute of Technology, Harbin 150001, China; leslieding@stu.hit.edu.cn (L.D.); shuaiyong@hit.edu.cn (Y.S.); libx@hit.edu.cn (B.L.)
[2] Aerospace Haiying (Harbin) Titanium Co., Ltd., Harbin 150001, China; hanfangming1@live.cn
[3] Department of Energy Sciences, Lund University, SE 22100 Lund, Sweden
* Correspondence: wangwei36@hit.edu.cn (W.W.); bengt.sunden@energy.lth.se (B.S.)

Abstract: In this work, a discontinuous airfoil fin printed circuit heat exchanger (PCHE) was used as a recuperator in a micro gas turbine system. The effects of the airfoil fin geometry parameters (arc height, maximum arc height position, and airfoil thickness) and the airfoil fin arrangements (horizontal and vertical spacings) on the PCHE channel's thermo-hydraulic performance were extensively examined by a numerical parametric study. The flow features, local heat transfer coefficient, and wall shear stress were examined in detail to obtain an enhanced heat transfer mechanism for a better PCHE design. The results show that the heat transfer and flow resistance were mainly increased at the airfoil leading edge owing to a flow jet, whereas the airfoil trailing edge had little effect on the thermo-hydraulic performance. The airfoil thickness was the most significant while the arc height and the vertical spacing were moderately significant to the performance. Moreover, only the airfoil thickness had a significant effect on the PCHE compactness. Based on a comprehensive investigation, two solutions NACA-6230 and -3220 were selected owing to their better thermal performance and smaller pressure drop, respectively, with horizontal spacings of 2 mm and vertical spacings of 2 or 3 mm.

Keywords: printed circuit heat exchanger; NACA airfoil fins; micro gas turbine recuperator; thermo-hydraulic; parametric study

Citation: Wang, W.; Ding, L.; Han, F.; Shuai, Y.; Li, B.; Sunden, B. Parametric Study on Thermo-Hydraulic Performance of NACA Airfoil Fin PCHEs Channels. *Energies* **2022**, *15*, 5095. https://doi.org/10.3390/en15145095

Academic Editor: Andrea De Pascale

Received: 15 June 2022
Accepted: 5 July 2022
Published: 12 July 2022

1. Introduction

A printed circuit heat exchanger (PCHE) as a new type of heat exchanger is widely used in high-pressure and high-temperature applications [1]. It possesses a high heat exchange area density (2500 m^2/m^3), high compactness, and a large heat transfer coefficient [2], due to the millimeter-level micro channels. Compared to some special finned surfaces [3,4] and metal foam heat exchangers [5], the unique diffusion welding technology of PCHE can improve its reliability. Because of the above advantages, PCHEs are commonly used in fuel gas supply systems, LNG regasification, and cryogenic applications, supercritical carbon dioxide (sCO_2) systems, and liquefied hydrogen and hydrogen refueling stations.

Apart from the above benefits of PCHE, a key problem of flow resistance needs to be solved when it is used in thermal power cycles. Numerous extensive studies have been conducted on this problem. Four types of PCHE channels have been the most studied [2,6], which are shown in Figure 1. Among these, the straight [7,8] and zigzag [9,10] fins are continuous fin types, whereas the S-type [11] and airfoil fins [12] are discontinuous fin types. Tsuzuki et al. [11] first found that an S-type fin channel has one-fifth the pressure drop of a zigzag channel at an equal heat transfer performance. Kim et al. [12] first designed an airfoil fin PCHE channel using the NACA-0020 model. The simulation results showed that the heat transfer coefficients of the airfoil fin and zigzag channels were similar; however, the pressure drop of the airfoil fin channel was reduced to one-twentieth of that of the zigzag channel.

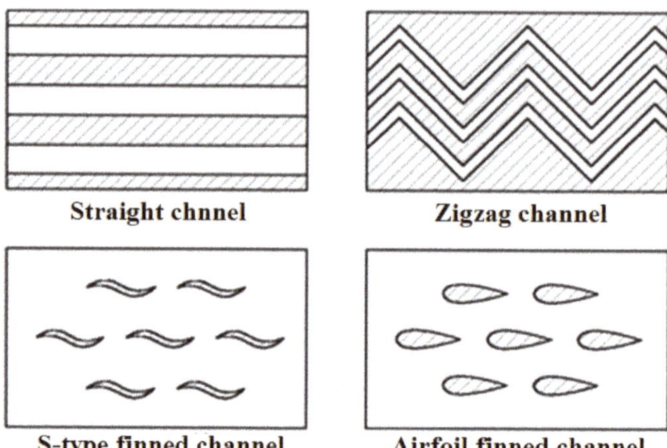

Figure 1. Four kinds of PCHE channels.

Because airfoil fin channels have demonstrated the best thermo-hydraulic performances, many studies have been conducted on airfoil fin PCHE applications in various fields. An experimental study of airfoil fin (NACA-0025) PCHE in a concentrating solar power system was performed by Wang et al. [13], with molten salt as a working fluid. The experimental data presented that the heat transfer performance of the airfoil fin PCHE was superior to the traditional PCHEs with straight and zigzag fins. Chen et al. [14] compared a NACA-00XX airfoil fin PCHE with a conventional zigzag PCHE used in the supercritical carbon dioxide Brayton cycle. The numerical results also presented that the pressure drop of the airfoil fin PCHE was remarkably reduced, and it maintained excellent heat transfer performance. The NACA-0010 airfoil fin PCHE demonstrated better comprehensive performance than the other three PCHEs (0020, 0030, and 0040). Xu et al. [15] performed a numerical study on four discontinuous fin configurations (rectangle, rounded rectangle, ellipse, and airfoil fins) in parallel and staggered arrangements. The results showed that the fin configurations had little effect on the overall thermo-hydraulic performances at low mass flow rates, and the airfoil fin with the staggered arrangement was better than the other types of fins. Chu et al. [16] studied three types of airfoil fin (NACA-8315, 8515, and 8715) PCHEs with consistent and reverse layouts used as condensers in the supercritical CO_2 Brayton power cycle. The results showed that the NACA-8515 airfoil fins with both consistent and reverse layouts on average improved the heat transfer coefficient by 28% and 11% with an increase of the pressure drop by 150% and 22%, respectively, compared with the symmetrical airfoil fins (NACA-0015). Ma et al. [17] employed an airfoil fin PCHE in a very high-temperature reactor and studied the effect of the fin-endwall fillet on the thermo-hydraulic performance. They found that the fin-endwall fillet increased the heat transfer and pressure drop with a longitudinal pitch of 1.63, whereas it had little effect when the longitudinal pitch was above 1.88. Although the above studies have examined airfoil fin PCHEs from different aspects, there is no report on the numerous parameters of airfoil fins and their arrangements. In addition, the enhanced heat transfer mechanism of airfoil fins needs to be clarified to guide PCHE design.

In this work, the discontinuous airfoil fin PCHE was employed as a micro gas turbine recuperator in the extended-range electric vehicle system [18], which has high requirements on heat transfer, pressure drop, and compactness, simultaneously. Therefore, the effects of the airfoil fin parameters (arc height, maximum arc height position, and airfoil thickness) and the airfoil fin arrangements (horizontal and vertical spacings) on the heat transfer and pressure drop were extensively examined by a parametric study. The flow features, local heat transfer coefficient, and wall shear stress were investigated to obtain an enhanced heat transfer mechanism for a better PCHE design.

2. Mathematical Approach

2.1. Physical Model and Boundary Conditions

In this work, the heat transfer and pressure drop of PCHE channels with different NACA airfoil fins were investigated. The structures of the two types of channels are shown in Figure 2. The examined geometrical parameters of the airfoil fin PCHE channels include the horizontal spacing (*Lh*), vertical spacing (*Lv*), airfoil chord (*Ll*), arc height (*hl*), maximum arc height position (*ll*), and airfoil thickness (*tl*), corresponding to the NACA airfoil rules.

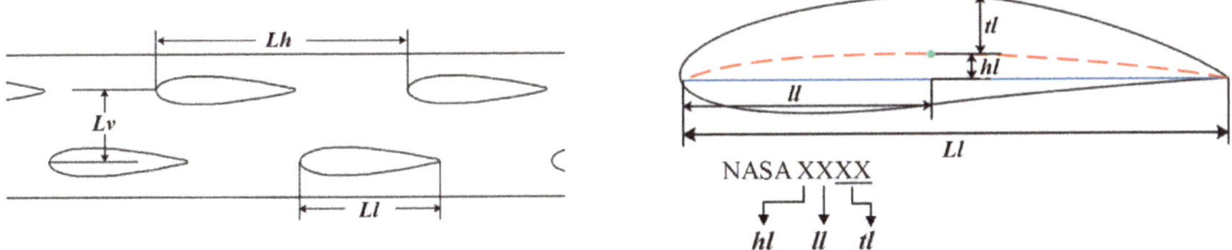

Figure 2. Structure parameters of NACA airfoil fins and arrangements.

The computational domain and boundary conditions are shown in Figure 3. Specifically, the inlet is a velocity boundary with a fixed temperature (T_{in} = 630 K), the outlet is a pressure boundary (P_{out} = 0.3 MPa), the left and right walls are periodic boundaries, and the up, down, and airfoil walls are fixed temperature (T_{wall} = 650 K) walls. The working fluid (compressed air) was incompressible, and its thermo-properties only varied with temperature, as can be seen from Table 1.

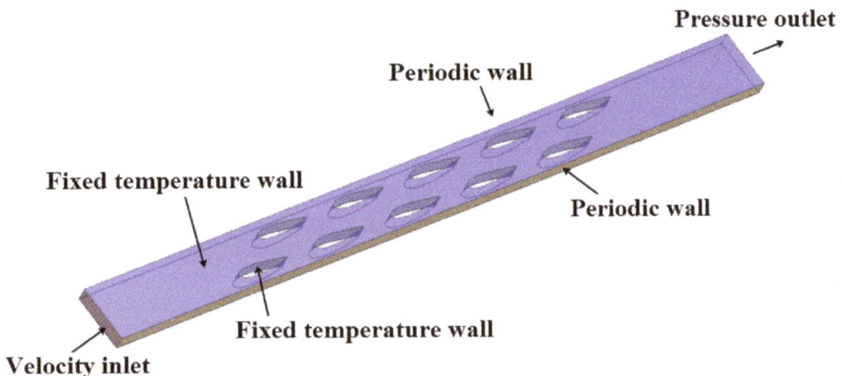

Figure 3. Computational domain and boundary conditions of airfoil fin channels.

Table 1. Thermo-physical properties of cold air (0.3 MPa, 600–720 K) [19].

Equation: $f(T)=c_1 \cdot T^2+c_2 \cdot T+c_3$				
	ρ (kg/m^3)	C_p (J/kg/K)	λ (W/m/K)	μ (kg/m/s)
c_1	4.801×10^{-6}	4.46×10^{-5}	-1.432×10^{-8}	-1.235×10^{-11}
c_2	-0.009275	0.1777	7.462×10^{-5}	5.021×10^{-8}
c_3	5.969	929.5	0.005678	5.191×10^{-6}

The structured grids were used for the airfoil fin channels' computational domain. The near wall grids were refined with y plus less than 1 [20], as shown in Figure 4. A grid independence test was performed on the NACA-0320 airfoil fin channel, as shown in Figure 5, and the total number of grids was chosen as 892,000.

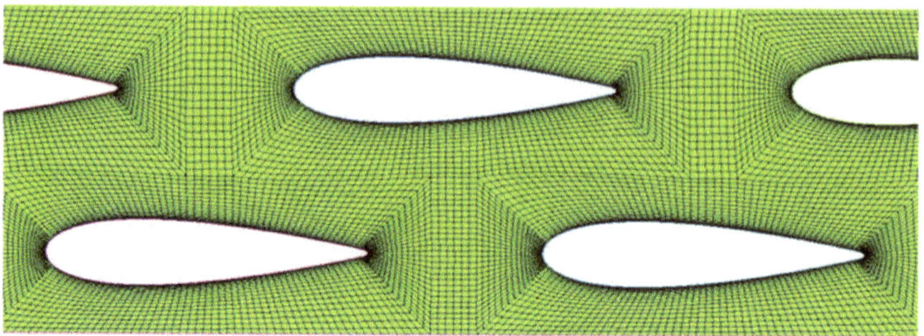

Figure 4. Computational grids of airfoil fin PCHE channels.

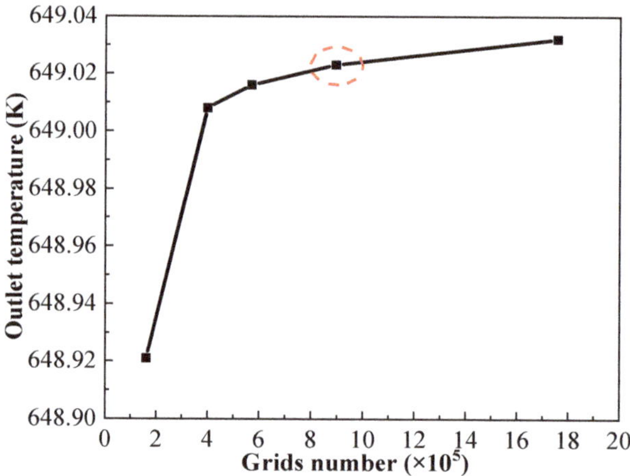

Figure 5. Grid independence test results for the PCHE channel with NACA-0320 airfoil fins.

2.2. Governing Equations and Numerical Approach

In this study, the realizable k-ε model was employed [21,22]; the governing equations are given below [23].

The continuity equation is

$$\frac{\partial(\rho u_i)}{\partial x_i} = 0 \tag{1}$$

The momentum equation is

$$\frac{\partial}{\partial x_j}(\rho u_i u_j) = -\frac{\partial P}{\partial x_i} + \frac{\partial}{\partial x_j}[\mu(\frac{\partial u_i}{\partial x_j} + \frac{\partial u_j}{\partial x_i} - \frac{2}{3}\delta_{ij}\frac{\partial u_j}{\partial x_j})] + \frac{\partial}{\partial x_j}(-\rho\overline{u_i' u_j'}) \tag{2}$$

The energy equation is

$$\frac{\partial}{\partial x_i}[u_i(\rho E + P)] = \frac{\partial}{\partial x_i}[(\lambda + \frac{c_P \mu_t}{\mathrm{Pr}_t})\frac{\partial T}{\partial x_j} + \mu u_i(\frac{\partial u_i}{\partial x_j} + \frac{\partial u_j}{\partial x_i} - \frac{2}{3}\delta_{ij}\frac{\partial u_j}{\partial x_j})] \tag{3}$$

where E is the total energy, $E = C_P T - P/\rho + u^2/2$, and λ is the thermal conductivity.

The k equation is

$$\frac{\partial}{\partial x_j}(\rho k u_j) = \frac{\partial}{\partial x_j}[(\mu + \frac{\mu_t}{\sigma_k})\frac{\partial k}{\partial x_j}] + G_k - \rho\varepsilon \tag{4}$$

The ε equation is

$$\frac{\partial}{\partial x_j}(\rho \varepsilon u_j) = \frac{\partial}{\partial x_j}[(\mu + \frac{\mu_t}{\sigma_\varepsilon})\frac{\partial \varepsilon}{\partial x_j}] + \rho C_1 S_\varepsilon - \rho C_2 \frac{\varepsilon^2}{k + \sqrt{v\varepsilon}} \tag{5}$$

where G_k represents the production of turbulent kinetic energy and is modeled as $G_k = \mu_t S^2$. The μ_t represents the eddy viscosity and is modeled as $\mu_t = \rho C_\mu k^2/\varepsilon$ [24].

In the above expressions, $C_1 = \max\left[0.43, \frac{\eta}{\eta+5}\right]$, $S = \sqrt{2S_{ij}S_{ij}}$, $S_{ij} = \frac{1}{2}\left(\frac{\partial u_i}{\partial x_j} + \frac{\partial u_j}{\partial x_i}\right)$, $C_2 = 1.9$, $\sigma_k = 1$, and $\sigma_\varepsilon = 1.2$ [25].

The semi-implicit method for the pressure-linked equations (*SIMPLE*) algorithm was employed on the velocity and pressure equations, and a second-order upwind scheme was employed on the energy and momentum equations. The scaled residuals for all solutions must be less than 10^{-6} [26,27] to obtain convergence.

2.3. Data Reduction

The average convective heat transfer coefficient (h), average pressure drop (ΔP_L), and total average heat flux (Q_A) were employed to assess the heat transfer, flow resistance, and compactness of the PCHEs, respectively [28–30].

$$h = \frac{q}{T_{wall} - 0.5(T_{in} + T_{out})} \tag{6}$$

$$\Delta P_L = \Delta P / L \tag{7}$$

$$Q_A = Q/A \tag{8}$$

where q is the average heat flux, L is the channel length, ΔP is the pressure drop between the inlet and outlet, and A is the area of heat exchange.

2.4. Numerical Model Validation

An experimental work by Ishizuka et al. [31] was used to validate the numerical model in this work. The experimental PCHE is a typically zigzag-type channel with the following geometrical parameters: hot-side channel pitch pl = 9 mm, angle α = 115°, and radius r = 0.95 mm; cold-side channel pitch pl = 7.24 mm, angle α = 100°, and radius r = 0.9 mm. The fluids for the two sides were both supercritical CO_2. The hot-side working fluid is supercritical CO_2; inlet temperature = 553.05 K, inlet pressure = 2.52 MPa, and mass flow rate = 0.867 g/s. The cold-side working fluid is also supercritical CO_2; inlet temperature = 381.05 K, inlet pressure = 8.28 MPa, and mass flow rate = 0.9456 g/s. The comparison of experimental data and the numerical results are shown in Table 2. The maximum error for the cold side is less than ±5%, and the maximum error for the hot side is less than ±1%, which meet the accuracy requirements.

Table 2. Numerical model validation using experimental data [31].

	Experimental	Numerical	Error
Pressure drop of cold side (Pa)	73,220	76,832.0	−3612 (−4.9%)
Pressure drop of hot side (Pa)	24,180	24,381.6	−2016.6 (−0.83%)
Temperature difference of cold side (K)	140.38	146.8	−6.42 (−4.6%)
Temperature difference of hot side (K)	169.6	171.3	−1.7 (−1.0%)

3. Results and Discussion

3.1. Effect of Arc Height (hl)

Four different arc heights (0330, 3330, 6330, and 9330) are compared in Figure 6. With the increase in arc height, the airfoils present more asymmetry, and the convective heat transfer coefficient (h), pressure drop per unit length (ΔP_L), and heat flux per unit area (Q_A) increase but to different extents. The average h values for 3330, 6330, and 9330 are 2.3%, 6.4%, and 8% higher than that for 0330, respectively. In addition, the average ΔP_L values for 3330, 6330, and 9330 are 3.3%, 15.4%, and 27.7% higher than that for 0330, respectively. The above trends suggest that the pressure drop increases rapidly with the increase in hl, whereas the heat transfer performance increase is moderate. Moreover, the Q_A increases slightly with increasing arc height (hl); this means that the hl is dis-significant to the PCHE compactness design.

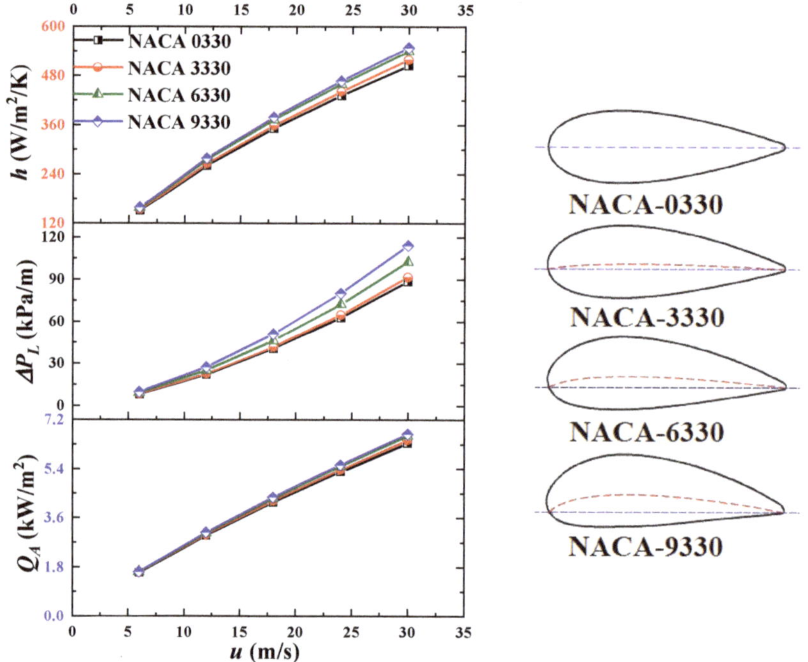

Figure 6. Average performance for different airfoil arc heights (hl).

The distributions of the velocity u, turbulent kinetic energy (TKE), local convective heat transfer coefficient (h_{local}), and wall shear stress (WSS) for NACA-0330 and 9330 are shown in Figure 7. From Figure 7a, velocity u for NACA-0330 is accelerated significantly in the main flow region, whereas for NACA-9330, this occurs near the airfoil surface, which is more beneficial for heat transfer improvement. The TKE for NACA-0330 is remarkable at the airfoil trailing edge, which indicates severe turbulent pulsations (see Figure 7b). The TKE for NACA-9330 is inhibited by the vertical velocity component. Figure 7c shows that the h_{local} reaching the maximum at the airfoil leading edge is due to the jet flow heat transfer being significant (than the convective heat transfer). The h_{local} reach to the minimum at the airfoil trailing edge is due to the secondary flow velocity being very low, causing a thick boundary layer. Moreover, the h_{local} for NACA-0330 in the main flow region and at the airfoil trailing edge is higher than that for NACA-9330, whereas the h_{local} for NACA-9330 in the near-wall region is significantly higher than that for NACA-0330. In Figure 7d, the WSS reflects the flow resistance created by the airfoil fins. It can be seen that the WSS is significant owing to the turbulent flow impingement on the airfoil leading edge, whereas it is minimum at the airfoil trailing edge owing to a secondary vortex generation. In addition, the WSS for NACA-9330 is overall greater than that for NACA-0330, based on the more drastic impact effect and the lower secondary flow generation of the former.

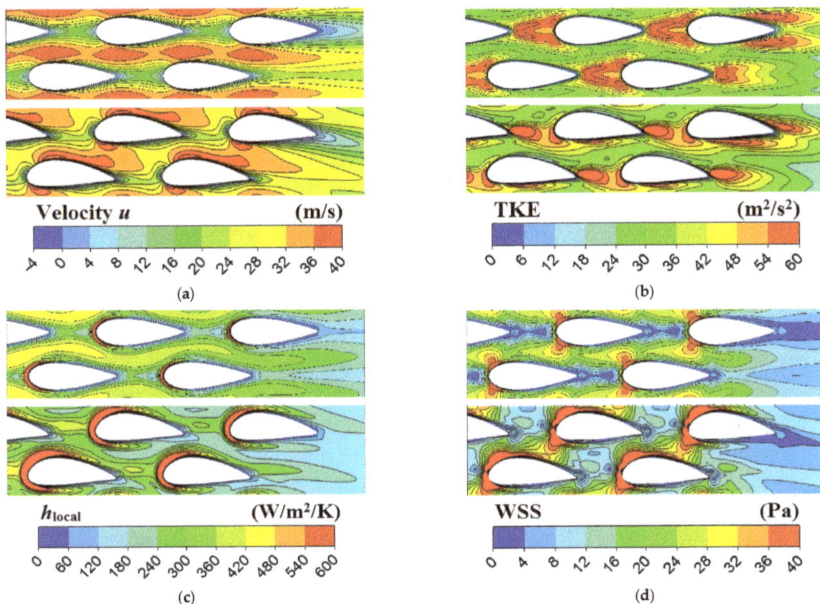

Figure 7. Contours of velocity u, TKE, h_{local}, and WSS for NACA-0330 and -9330 airfoils. (**a**) Velocity u distributions for NACA 0330 and 9330 (middle surface, u_{in} = 18 m/s). (**b**) TKE distributions for NACA 0330 and 9330 (middle surface, u_{in} = 18 m/s). (**c**) h_{local} distributions for NACA 0330 and 9330 (wall surface, u_{in} = 18 m/s). (**d**) WSS distributions for NACA 0330 and 9330 (wall surface, u_{in} = 18 m/s).

3.2. *Effect of Maximum Arc Height Position (ll)*

The maximum arc height position (ll) mainly affects the second half of the airfoil structure, whereas it has a minor effect on the first half, as in the cases of NACA-6220, 6420, 6620, and 6820 shown in Figure 8. The average h, ΔP_L, and Q_A values for the four kinds of ll indicate that the ll is not significant to the three evaluation performances. Relatively, h and ΔP_L for $ll < 5$ (NACA-6220 and -6420) cause slightly greater effects than that for $ll > 5$ (NACA-6620 and -6820). Moreover, the Q_A values for the four cases are similar.

The distribution of the velocity u, TKE, h_{local}, and WSS for NACA-6220 and -6820 are shown in Figure 9. The velocity u for NACA-6220 presents a high acceleration at the airfoil leading edge, whereas it exhibits a low back flow at the airfoil trailing edge. The TKE for NACA-6820 is significantly higher than that for NACA-6220 after the airfoil trailing edge. The above flow feature causes the h_{local} for NACA-6220 to be much higher at the airfoil leading edge than that for NACA-6280. In comparison, the h_{local} for NACA-6820 is much higher in the main flow region than that for NACA-6220 owing to the high TKE distribution. The WSSs also present similar tendencies as the h_{local} distributions for NACA-6220 and -6820, which are also determined by the flow features. In comparison, the partial differences between the h_{local} and the WSS do not cause remarkable variations in the overall performance, as shown in Figure 8.

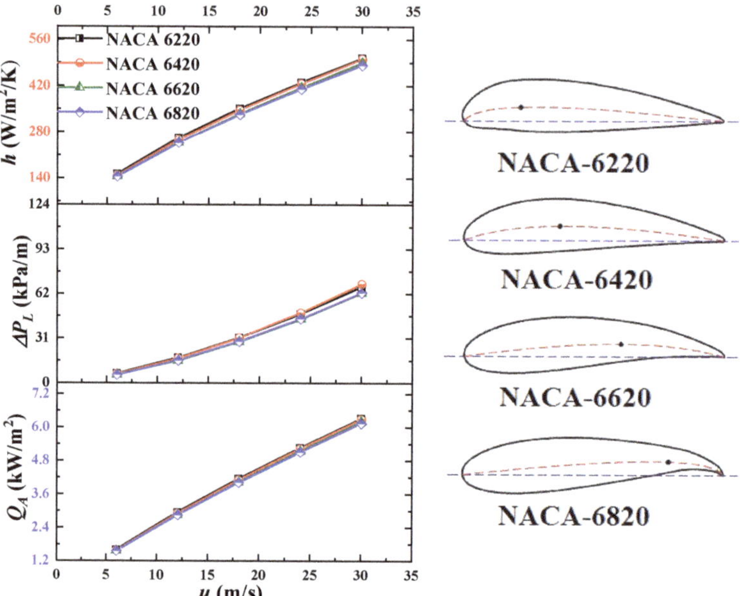

Figure 8. Average performances for different locations of maximum arc height (*ll*).

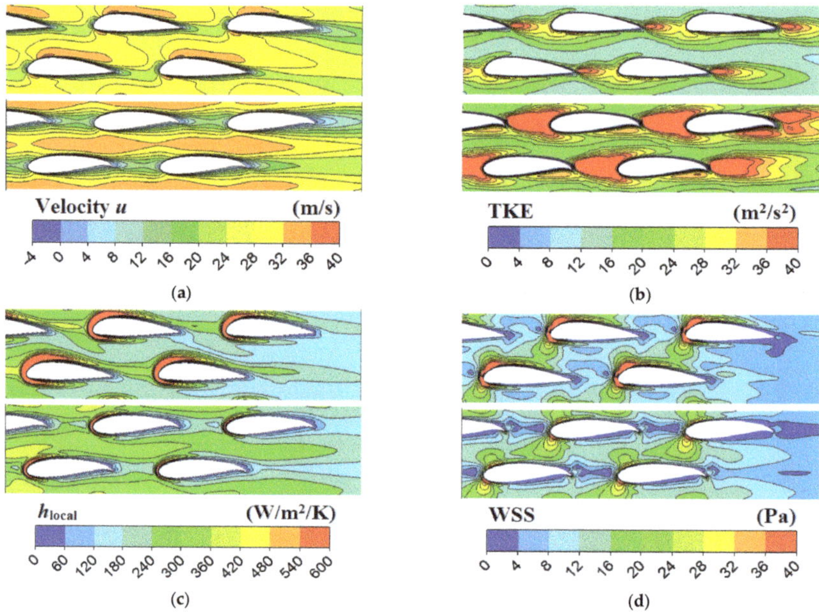

Figure 9. Contours of velocity u, TKE, h_{local}, and WSS for NACA-6220 and 6820 airfoils. (**a**) Velocity u distributions for NACA 6220 and 6820 (middle surface, u_{in} = 18 m/s). (**b**) TKE distributions for NACA 6220 and 6820 (middle surface, u_{in} = 18 m/s). (**c**) h_{local} distributions for NACA 6220 and 6820 (wall surface, u_{in} = 18 m/s). (**d**) WSS distributions for NACA 6220 and 6820 (wall surface, u_{in} = 18 m/s).

3.3. Effect of Airfoil Thickness (tl)

The effect of the airfoil thickness (*tl*) on the average performance is shown in Figure 10. The structure diagrams show significant differences among the four cases. The width increases with the increase in *tl*, and the average h, ΔP_L, and Q_A are all obviously heightened with the increase in *tl*. The average h for 0320, 0330, and 0340 are 9.58%, 21.4%, and 33.8% higher than that of 0310, respectively. The average ΔP_L for 0320, 0330, and 0340 are 38.1%, 113%, and 246.25% higher than that of 0310, respectively. Additionally, the increase in airfoil thickness is an advantage to improve the PCHE compactness.

The distributions of the velocity u, TKE, h_{local}, and WSS for NACA-0310 and 0340 are shown in Figure 11. The velocity u for NACA-0310 shows a relatively uniform distribution, owing to the small distribution of the narrow airfoil. In comparison, the velocity u for NACA-0340 presents a severe acceleration between two rows of the airfoils, owing to the reduction in the cross-section area by the large width of the airfoils. In addition, the velocity u distribution shown in the red dotted frame in Figure 11a indicates that although the large *tl* causes a high-intensity reverse flow at the airfoil trailing edge, it is inhibited by the accelerating fluid. NACA-0340 has an extremely higher TKE than NACA-0310 and, partially, the TKE can be up to 10 times higher. The h_{local} and WSS for NACA-0340 present significant increases at the airfoil leading edge because the increase in the windward area causes a remarkable jet flow effect. In comparison, the h_{local} and NACA-WSS for 0340 are slightly increased by the interstitial flows.

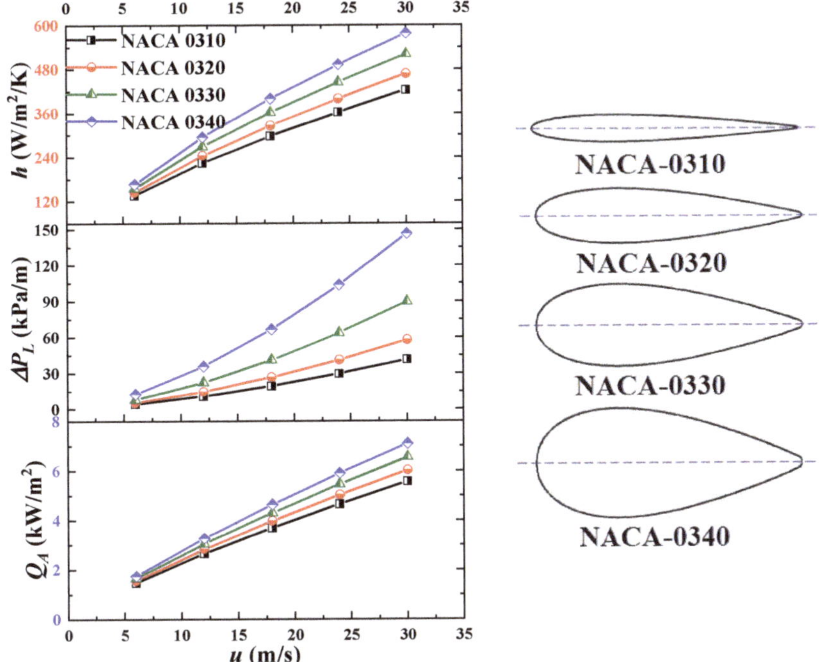

Figure 10. Average performance for different airfoil thicknesses (*tl*).

3.4. Effects of Horizontal Spacing (Lh) and Vertical Spacing (Lv)

The average performance for different horizontal spacing (*Lh*) is shown in Figure 12. The h, ΔP_L, and Q_A slightly decrease with the increase in *Lh*. The average h values of $Lh = 3$, 4, and 5 mm are 2.2%, 4.1%, and 6.4% lower than that of $Lh = 2$ mm, respectively. The average ΔP_L of $Lh = 3$, 4, and 5 mm are 6.2%, 9.6%, and 12.2% lower than that of $Lh = 2$ mm, respectively. From Figure 13, it can be seen that the h_{local} and WSS distributions for two cases ($Lh = 2$ and 5 mm) are similar around the airfoil walls. This suggests that the variation in Lh has no effect on the enhanced heat transfer mechanism, and the variation rate of the average performance is uniform. This is because of the secondary flow at each airfoil trailing edge, which leads to a periodic flow redevelopment.

The average performance for different vertical spacings (Lv) is shown in Figure 14. The average h of Lv = 3, 4, and 5 mm are 4.2%, 6.4%, and 6.6% lower than that of Lv = 2 mm. The average ΔP_L of Lv = 3, 4, and 5 mm are 26.3%, 37.1%, and 42% lower than that of Lv = 2 mm, respectively. The average h and ΔP_L show that the performances of Lv = 4, and 5 mm are quite propinquity. It means that when Lv is above 3 mm, the vortices caused by the airfoils are not coupled to each other anymore. Moreover, the Lv > 3 mm is not suggested due to the compactness demands. From a comprehensive point, the Lv = 3 mm is suggested due to the average h being smaller—lower than that of Lv = 2 mm, and the average ΔP_L is significantly lower than that of Lv = 2 mm. The global distributions of the velocity u, TKE, h_{local}, and WSS for Lv = 2 mm in Figure 15, also show that the disturbed flow and turbulent pulsation between the two rows of the airfoils are superimposed, fundamentally causing the heat transfer and flow resistance to increase. In comparison, the corresponding distribution for Lv = 5 mm suggests that the complex flow features caused by the airfoils are independent of each other.

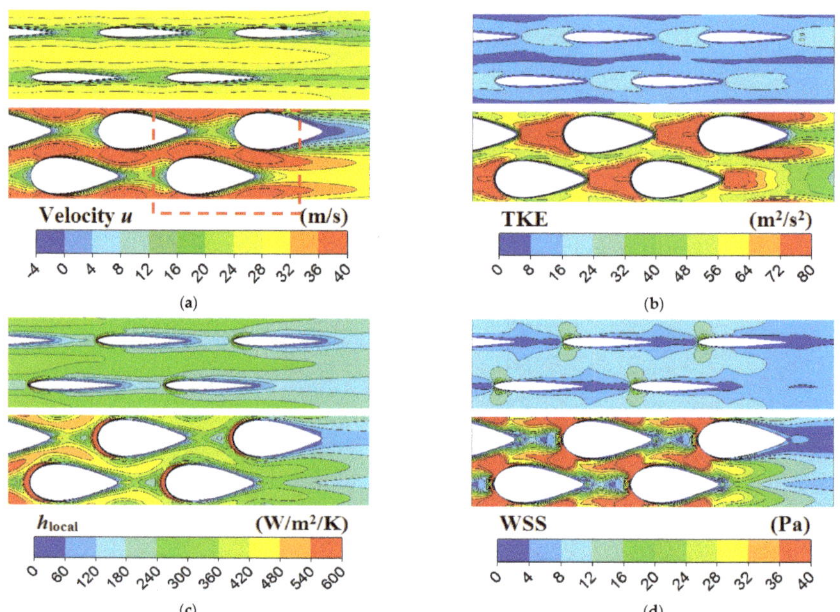

Figure 11. Contours of velocity u, TKE, h_{local}, and WSS for NACA-0310 and 0340 airfoils. (**a**) Velocity u distributions for NACA 0310 and 0340 (middle surface, u_{in} = 18 m/s). (**b**) TKE distributions for NACA 0310 and 0340 (middle surface, u_{in} = 18 m/s). (**c**) h_{local} distributions for NACA 0310 and 0340 (wall surface, u_{in} = 18 m/s). (**d**) WSS distributions for NACA 0310 and 0340 (wall surface, u_{in} = 18 m/s).

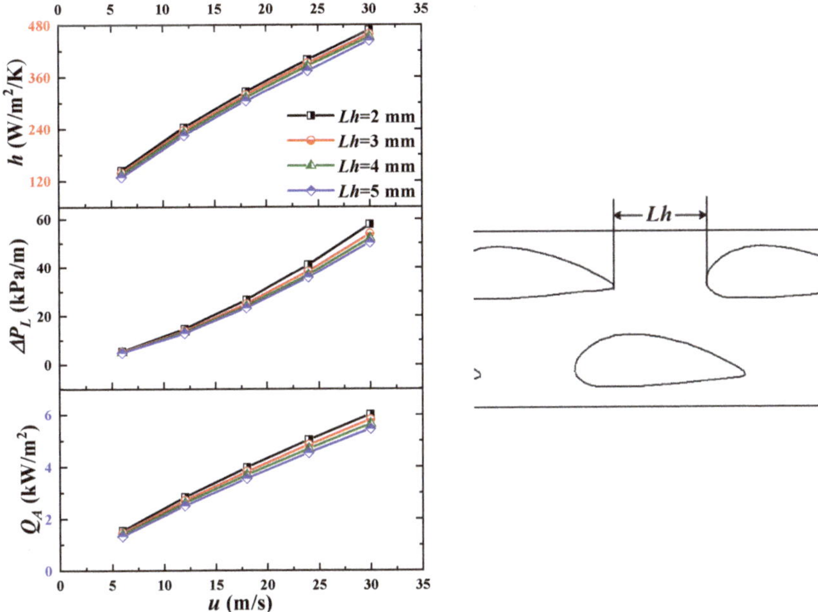

Figure 12. Average performance for different horizontal spacings (*Lh*).

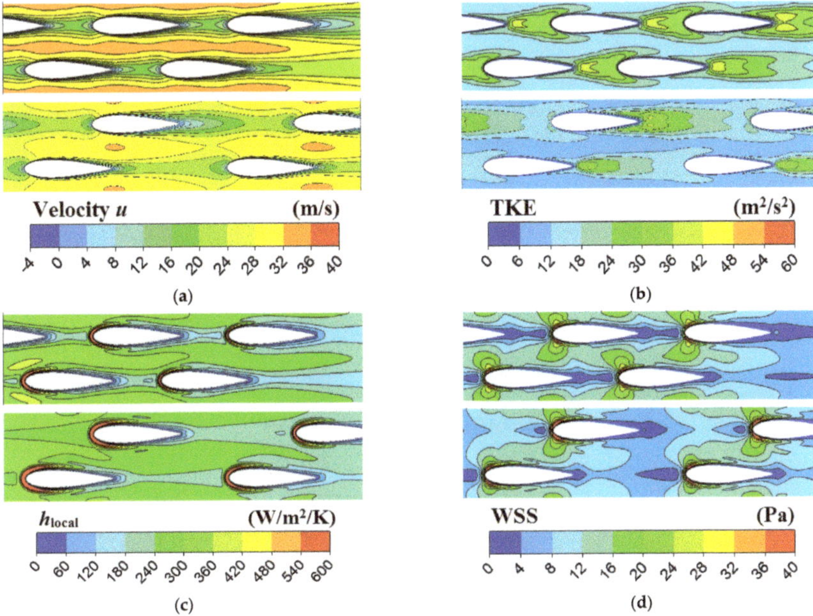

Figure 13. Contours of velocity u, TKE, h_{local}, and WSS for Lh = 2 and 5 mm with NACA-0320. (**a**) Velocity u distributions for Lh = 2 and 5 mm (0320, middle surface, u_{in} = 18 m/s). (**b**) TKE distributions for Lh = 2 and 5 mm (0320, middle surface, u_{in} = 18 m/s). (**c**) h_{local} distributions for Lh = 2 and 5 mm (0320, wall surface, u_{in} = 18 m/s). (**d**) WSS distributions for Lh = 2 and 5 mm (0320, wall surface, u_{in} = 18 m/s).

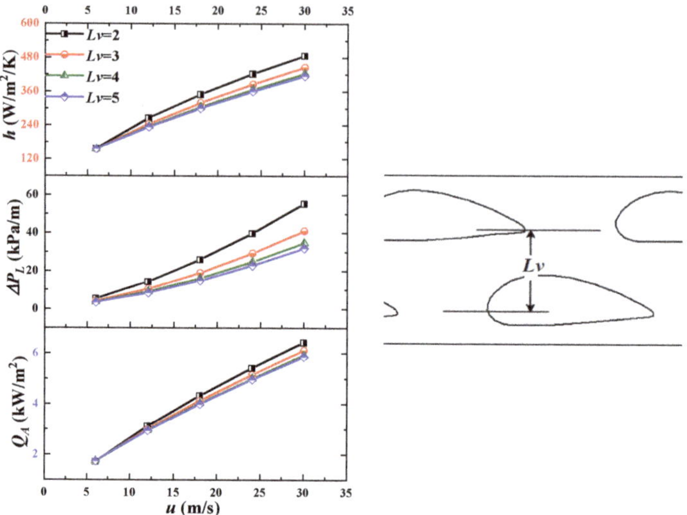

Figure 14. Average performance for different vertical spacings (Lv).

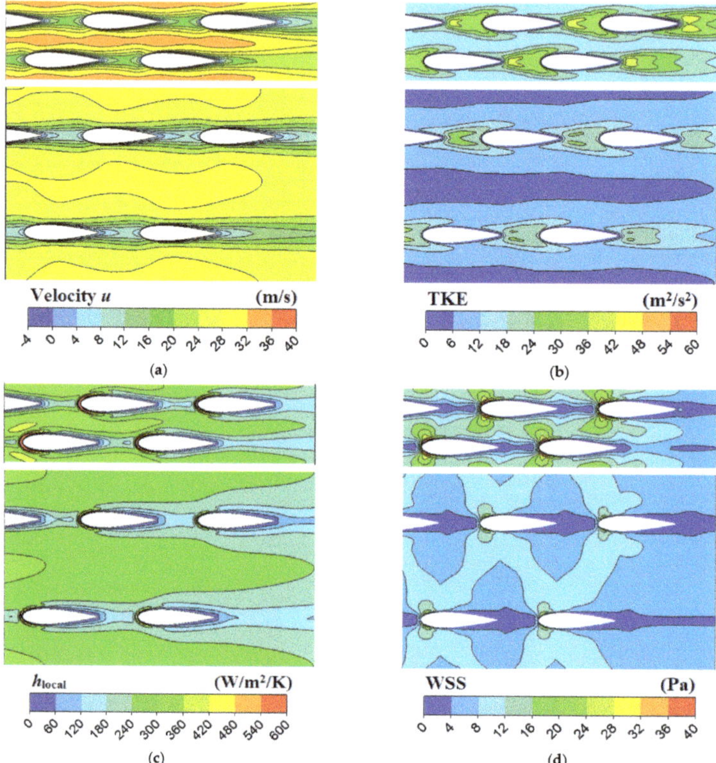

Figure 15. Contours of velocity u, TKE, h_{local}, and WSS for Lv = 2 and 5 mm with NACA-0320. (**a**) Velocity u distributions for Lv = 2 and 5 mm (0320, middle surface, u_{in} = 18 m/s). (**b**) TKE distributions for Lv = 2 and 5 mm (0320, middle surface, u_{in} = 18 m/s). (**c**) h_{local} distributions for Lv = 2 and 5 mm (0320, wall surface, u_{in} = 18 m/s). (**d**) WSS distributions for Lv = 2 and 5 mm (0320, wall surface, u_{in} = 18 m/s).

4. Conclusions

In this study, airfoil fins PCHEs were employed as micro gas turbine recuperators for extended-range electric vehicles, and the thermo-hydraulic performances of PCHE channels with airfoil fins were numerically studied. Five geometric parameters, i.e., arc height (*hl*), maximum arc height position (*ll*), airfoil thickness (*tl*) horizontal spacing (*Lh*), and vertical spacing (*Lv*) were examined in detail, and the enhanced heat transfer mechanism was analyzed. The main conclusions are summarized as follows:

(1) The heat transfer and the flow resistance are mainly increased at the airfoil leading edge owing to the flow jet. A secondary vortex is produced at the airfoil trailing edge, causing a violent turbulent pulsation; however, it has a minor effect on the heat transfer improvement.

(2) Among the five geometrical parameters, the airfoil thickness (*tl*) is the most significant. The arc height (*hl*) and the vertical spacing (*Lv*) are moderately significant, whereas the maximum arc height position (*ll*) and the horizontal spacing (*Lh*) are almost insignificant to the thermo-hydraulic performance. Moreover, only the airfoil thickness has a significant effect on the PCHE compactness.

(3) Two solutions (NACA-6230 and -3220) were selected for their better thermal performance and smaller pressure drop, respectively, with Lh = 2 mm and Lv = 2 or 3 mm.

Author Contributions: Conceptualization, Y.S.; Funding acquisition, W.W.; Investigation, L.D.; Methodology, F.H.; Resources, B.L.; Writing—original draft, W.W.; Writing—review & editing, B.S. All authors have read and agreed to the published version of the manuscript.

Funding: National Natural Science Foundation of China (52106082), and Heilongjiang Province Science and Technology Major Project (2020ZX10A03).

Acknowledgments: This study is supported by the National Natural Science Foundation of China (52106082), China Postdoctoral Science Foundation (2019M661279), Heilongjiang Provincial Postdoctoral Science Foundation (LBH-Z19162), and Heilongjiang Province Science and Technology Major Project (2020ZX10A03).

Conflicts of Interest: The authors declare that they have no known competing financial interests or personal relationships that could have appeared to influence the work reported in this paper.

Nomenclature

A	Area (m^2)
C_1, C_2	Realizable k-ε model constants
Cp	Specific heat (J/kg/K)
E	Internal energy (J/kg)
h	Heat transfer coefficient ($W/m^2/K$)
L	Length (m)
Lh, Lv	Horizontal and vertical spacing (mm)
Ll, hl, ll, tl	Airfoil chord, arc height, maximum are height position and airfoil thickness
T	Temperature (K)
TKE	Turbulent kinetic energy (J/kg)
P	Pressure (Pa)
PCHE	Printed circuit heat exchanger
Pr	Prandtl number
Q	Heat flux (W)
u, v, w	Streamwise, transverse, and vertical velocity components (m/s)
WSS	Wall shear stress

Greek letters	
ε	Turbulence dissipation rate (m^3/s^2)
λ	Thermal conductivity (W/m/K)
μ	Dynamic viscosity (kg/m/s)
ρ	Density of fluid (kg/m^3)
Subscripts	
in, out	Inlet and outlet
i, j, k	Directions of the coordinate system
wall	Wall

References

1. Chen, M.; Sun, X.; Christensen, R.N.; Shi, S.; Skavdahl, I.; Utgikar, V.; Sabharwall, P. Experimental and numerical study of a printed circuit heat exchanger. *Ann. Nucl. Energy* **2016**, *97*, 221–231. [CrossRef]
2. Huang, C.; Cai, W.; Wang, Y.; Liu, Y.; Li, Q.; Li, B. Review on the characteristics of flow and heat transfer in printed circuit heat exchangers. *Appl. Therm. Eng.* **2019**, *153*, 190–205. [CrossRef]
3. Batista, J.; Trp, A.; Lenic, K. Heat Transfer Enhancement of Crossflow Air-to-Water Fin-and-Tube Heat Exchanger by Using Delta-Winglet Type Vortex Generators. *Energies* **2022**, *15*, 2070. [CrossRef]
4. Altwieb, M.; Mishra, R.; Aliyu, A.M.; Kubiak, K.J. Heat Transfer Enhancement by Perforated and Louvred Fin Heat Exchangers. *Energies* **2022**, *15*, 400. [CrossRef]
5. Hmad, A.A.; Dukhan, N. Cooling Design for PEM Fuel-Cell Stacks Employing Air and Metal Foam: Simulation and Experiment. *Energies* **2021**, *14*, 2687. [CrossRef]
6. Yoon, S.H.; No, H.C.; Kang, G.B. Assessment of straight, zigzag, S-shape, and airfoil PCHEs for intermediate heat exchangers of HTGRs and SFRs. *Nucl. Eng. Des.* **2014**, *270*, 334–343. [CrossRef]
7. Chu, W.-X.; Li, X.-H.; Ma, T.; Chen, Y.-T.; Wang, Q.-W. Experimental investigation on SCO2-water heat transfer characteristics in a printed circuit heat exchanger with straight channels. *Int. J. Heat Mass Transf.* **2017**, *113*, 184–194. [CrossRef]
8. Jiang, Y.; Liese, E.; Zitney, S.E.; Bhattacharyya, D. Design and dynamic modeling of printed circuit heat exchangers for supercritical carbon dioxide Brayton power cycles. *Appl. Energy* **2018**, *231*, 1019–1032. [CrossRef]
9. Kim, J.H.; Baek, S.; Jeong, S.; Jung, J. Hydraulic performance of a microchannel PCHE. *Appl. Therm. Eng.* **2010**, *30*, 2157–2162. [CrossRef]
10. Meshram, A.; Jaiswal, A.K.; Khivsara, S.D.; Ortega, J.D.; Ho, C.; Bapat, R.; Dutta, P. Modeling and analysis of a printed circuit heat exchanger for supercritical CO2 power cycle applications. *Appl. Therm. Eng.* **2016**, *109*, 861–870. [CrossRef]
11. Tsuzuki, N.; Kato, Y.; Ishiduka, T. High performance printed circuit heat exchanger. *Appl. Therm. Eng.* **2007**, *27*, 1702–1707. [CrossRef]
12. Kim, D.E.; Kim, M.H.; Cha, J.E.; Kim, S.O. Numerical investigation on thermal–hydraulic performance of new printed circuit heat exchanger model. *Nucl. Eng. Des.* **2008**, *238*, 3269–3276. [CrossRef]
13. Wang, W.-Q.; Qiu, Y.; He, Y.-L.; Shi, H.-Y. Experimental study on the heat transfer performance of a molten-salt printed circuit heat exchanger with airfoil fins for concentrating solar power. *Int. J. Heat Mass Transf.* **2019**, *135*, 837–846. [CrossRef]
14. Chen, F.; Zhang, L.; Huai, X.; Li, J.; Zhang, H.; Liu, Z. Comprehensive performance comparison of airfoil fin PCHEs with NACA 00XX series airfoil. *Nucl. Eng. Des.* **2017**, *315*, 42–50. [CrossRef]
15. Xu, X.Y.; Wang, Q.W.; Li, L.; Ekkad, S.V.; Ma, T. Thermal-Hydraulic Performance of Different Discontinuous Fins Used in a Printed Circuit Heat Exchanger for Supercritical CO_2. *Numer. Heat Transf. Part A Appl.* **2015**, *68*, 1067–1086. [CrossRef]
16. Chu, W.-X.; Bennett, K.; Cheng, J.; Chen, Y.-T.; Wang, Q.-W. Thermo-Hydraulic Performance of Printed Circuit Heat Exchanger with Different Cambered Airfoil Fins. *Heat Transf. Eng.* **2019**, *41*, 708–722. [CrossRef]
17. Ma, T.; Xin, F.; Li, L.; Xu, X.-Y.; Chen, Y.-T.; Wang, Q.-W. Effect of fin-endwall fillet on thermal hydraulic performance of airfoil printed circuit heat exchanger. *Appl. Therm. Eng.* **2015**, *89*, 1087–1095. [CrossRef]
18. Xiao, G.; Yang, T.; Liu, H.; Ni, D.; Ferrari, M.L.; Li, M.; Luo, Z.; Cen, K.; Ni, M. Recuperators for micro gas turbines: A review. *Appl. Energy* **2017**, *197*, 83–99. [CrossRef]
19. Wang, W.; Li, B.; Tan, Y.; Li, B.; Shuai, Y. Multi-objective optimal design of NACA airfoil fin PCHE recuperator for micro-gas turbine systems. *Appl. Therm. Eng.* **2021**, *204*, 117864. [CrossRef]
20. Mauro, G.M.; Iasiello, M.; Bianco, N.; Chiu, W.K.S.; Naso, V. Mono- and Multi-Objective CFD Optimization of Graded Foam-Filled Channels. *Materials* **2022**, *15*, 968. [CrossRef]
21. Fernández, I.; Sedano, L. Design analysis of a lead–lithium/supercritical CO2 Printed Circuit Heat Exchanger for primary power recovery. *Fusion Eng. Des.* **2013**, *88*, 2427–2430. [CrossRef]
22. Wang, W.; Shuai, Y.; Li, B.; Li, B.; Lee, K.-S. Enhanced heat transfer performance for multi-tube heat exchangers with various tube arrangements. *Int. J. Heat Mass Transf.* **2021**, *168*, 120905. [CrossRef]
23. Fluent, A. *Ansys Fluent Theory Guide*; ANSYS Inc.: Canonsburg, PN, USA, 2011; Volume 15317, pp. 724–746.
24. Shih, T.-H.; Liou, W.W.; Shabbir, A.; Yang, Z.; Zhu, J. A new k-ϵ eddy viscosity model for high reynolds number turbulent flows. *Comput. Fluids* **1995**, *24*, 227–238. [CrossRef]

25. Versteeg, W.M.H.K. *An Introduction to Computational Fluid Dynamics: The Finite Volume Method*, 2nd ed.; Prentice Hall: Harlow, UK, 2007.
26. Wang, W.; Zhang, Y.; Li, B.; Han, H.; Gao, X. Influence of geometrical parameters on turbulent flow and heat transfer characteristics in outward helically corrugated tubes. *Energy Convers. Manag.* **2017**, *136*, 294–306. [CrossRef]
27. Feng, Y.-Y.; Wang, C.-H. Discontinuous finite element method applied to transient pure and coupled radiative heat transfer. *Int. Commun. Heat Mass Transf.* **2021**, *122*, 105156. [CrossRef]
28. Wang, W.; Shuai, Y.; Ding, L.; Li, B.; Sundén, B. Investigation of complex flow and heat transfer mechanism in multi-tube heat exchanger with different arrangement corrugated tube. *Int. J. Therm. Sci.* **2021**, *167*, 107010. [CrossRef]
29. NematpourKeshteli, A.; Iasiello, M.; Langella, G.; Bianco, N. Enhancing PCMs thermal conductivity: A comparison among porous metal foams, nanoparticles and finned surfaces in triplex tube heat exchangers. *Appl. Therm. Eng.* **2022**, *212*, 118623. [CrossRef]
30. Sun, K.; Zhou, G.-Y.; Luo, X.; Tu, S.-T.; Huang, Y.-Y. Study on enhanced heat transfer performance of open-cell metal foams based on a hexahedron model. *Numer. Heat Transfer Part A Appl.* **2022**, 1–21. [CrossRef]
31. Ishizuka, T. Thermal-Hydraulic Characteristics of a Printed Circuit Heat Exchanger in a Supercritical CO_2 Loop. In Proceedings of the 11th International Topical Meeting on Nuclear Reactor Thermal-Hydraulics, Avignon, France, 2–6 October 2005.

Article

Numerical Simulation of Ventilation Performance in Mushroom Solar Greenhouse Design

Yiming Li [1,2,3,4], Fujun Sun [1], Wenbin Shi [2,3,4,5], Xingan Liu [2,3,4,5,*] and Tianlai Li [2,3,4,5,*]

1 College of Engineering, Shenyang Agricultural University, No. 120 Dongling Road, Shenhe District, Shenyang 110866, China
2 National & Local Joint Engineering Research Center of Northern Horticultural Facilities Design & Application Technology (Liaoning), No. 120 Dongling Road, Shenhe District, Shenyang 110866, China
3 Modern Facility Horticulture Engineering Technology Center (Shenyang Agricultural University), No. 120 Dongling Road, Shenhe District, Shenyang 110866, China
4 Key Laboratory of Protected Horticulture (Shenyang Agricultural University), Ministry of Education, No. 120 Dongling Road, Shenhe District, Shenyang 110866, China
5 College of Horticulture, Shenyang Agricultural University, No. 120 Dongling Road, Shenhe District, Shenyang 110866, China
* Correspondence: lxa10157@syau.edu.cn (X.L.); ltl@syau.edu.cn (T.L.)

Abstract: Numerical simulation is an effective tool for the thermal management of propulsion systems. Moreover, it contributes to the design and performance assessment of solar greenhouses for mushroom ventilation. Because the planning and design of the clustered solar greenhouse are still undiscovered, this study has developed a 3-D mathematical model suitable for a large-scale park of mushroom solar greenhouses based on computational fluid dynamics (CFD) theory. The effects of the orientation arrangement, horizontal spacing, vertical spacing of the cultivation racks, and the building distance between adjacent greenhouses on the ventilation performance were analyzed. The numerical simulation showed good agreement with the experimental measurement. The CFD results indicated that the reasonable layout of cultivation racks in mushroom solar greenhouses is a north-south arrangement. The horizontal spacing of cultivation racks has a significant influence on the wind speed and cooling rate, and the optimal spacing is 0.8 m. The overall height of the cultivation racks has little effect on the ventilation performance. Nevertheless, the vertical spacing between cultivation rack layers has a remarkable effect, and the optimal vertical spacing is 0.29 m. Reducing the building distance between the two adjacent greenhouses within a certain range helps increase the ventilation efficiency, leading to an increase in land utilization in the greenhouse park. The optimal building distance between the adjacent greenhouses is 10 m. The research results can provide theoretical guidance for improving the production quality and land utilization of mushroom facilities.

Keywords: ventilation; solar greenhouse; numerical simulation; system design

Citation: Li, Y.; Sun, F.; Shi, W.; Liu, X.; Li, T. Numerical Simulation of Ventilation Performance in Mushroom Solar Greenhouse Design. *Energies* **2022**, *15*, 5899. https://doi.org/10.3390/en15165899

Academic Editors: Bengt Sunden, Jian Liu, Chaoyang Liu and Yiheng Tong

Received: 20 July 2022
Accepted: 12 August 2022
Published: 14 August 2022

1. Introduction

Propulsion systems are important, widespread, and reliable devices in many fields such as aircraft engines, rocket engines, and scramjets. To improve the thermal management and flow control, numerical simulation has been employed as an effective method for the design and performance assessment. Computational fluid dynamics (CFD) plays an important role in the numerical simulation and is widely used in the field of engineering design, such as for solar greenhouse [1–3]. The solar greenhouse is a unique and distinctive greenhouse type in northern China, which does not need to consume fuel for heating in the production process [4,5]. It is a type of high-efficiency and energy-saving horticultural facility [6]. Even in the cold season, the solar greenhouse can only rely on sunlight to meet the needs of crop growth due to the conversion of heat energy to maintain the indoor temperature [7–9]. The production and consumption of edible fungi in the world is very large [10]. In recent years, the planting area and output of mushrooms have shown a multiple growth. It is worth mentioning that China's annual output of mushrooms can reach more than 70% of the global total. This is mainly attributed to the rapid development of the special solar greenhouse for mushroom production [11]. Although the heat flux in the greenhouse mostly comes from solar radiation and energy storage during the day, heating systems are used to deal with extreme weather in the design of the greenhouse. Zheng et al. [12] investigated the thermal performance of various nanofluids in heat exchangers. The tested optimal Fe_3O_4-water nanofluid was used as a heat transfer medium in the heating system. The results provided an empirical formula of Nusselt numbers for adjustment strategies under different working conditions. The effects of various surfactants on the stability and thermophysical properties of nanofluids were investigated experimentally after the preparation of the nanofluids [13], and the thermal efficiencies of different nanofluids were analyzed in a plate heat exchanger [14], a pipe [15], and an indoor electric heater under a magnetic field [16]. In order to ensure the high yield and high quality of crops in solar greenhouses, it is necessary to properly control the main environmental factors, such as temperature, humidity, and CO_2 concentration [17,18]. The most practical and most effective way to control temperature is ventilation. Therefore, the planning and design of the mushroom solar greenhouse can be carried out according to ventilation performance. As mushrooms are placed on the cultivation rack in the form of fungus sticks, the layout of the cultivation racks is extremely important. However, the existing construction principle for a clustered greenhouse park is based on the effective light interception, which does not fit into the weak-light habit of mushroom.

The ventilation plays a very important role in the greenhouse management, directly affecting the yield of mushrooms. A reasonable ventilation design of the mushroom solar greenhouse can not only regulate the internal environment of the greenhouse, but can also improve the efficiency of energy utilization [19]. Firstly, the ventilation can eliminate excess heat inside the greenhouse to the outside environment, so as to reduce the internal humidity and harmful gases [20]. Secondly, due to the external and internal air circulation, the temperature distribution in the greenhouse will become more uniform. The elimination of extreme temperature helps adjust the micro-environment to avoid disease. Thirdly, the air exchange can also accelerate the airflow, which is conducive to the growth and development of mushrooms [21]. Bournet et al. [22] performed a CFD simulation to investigate the vent configuration in the glass greenhouse. Villagrán et al. [23] studied the effect of roof vents on the ventilation performance of multi-span greenhouses. He et al. [24] evaluated the ventilation characteristics of a plastic greenhouse. Bartzanas et al. [25] investigated the natural ventilation in an arched greenhouse. Pakari and Ghani [26] analyzed the airflow distribution in a greenhouse equipped with wind towers. Zhang et al. [27] assessed the impacts of vent structure in Chinese solar greenhouses. Nevertheless, there are few studies on the ventilation of mushroom solar greenhouses. The mushroom solar greenhouse has high requirements regarding temperature, humidity, and other environmental factors. In general, the research results of a vegetable greenhouse cannot be copied to mushroom solar greenhouses. Therefore, there is an urgent need to fill that gap.

Mushrooms are produced in the form of fungus sticks in cultivation racks. Therefore, the ventilation is more complicated than soil cultivation. The uneven airflow distribution may reduce mushroom growth speed, so the greenhouse ventilation for mushroom fungus sticks plays a significant role [28]. Although scholars have carried out some research on greenhouse ventilation, the reported literature mostly takes the empty greenhouse as the research object. Han et al. [29] developed a ventilation equation for a mushroom greenhouse based on functional heat balance, and the air vents were designed to improve the temperature distribution. Zhang et al. [30] proposed a method to install ventilation pipes on the cultivation racks to improve the ventilation performance. Beak et al. [31] used CFD to simulate the effect of airflow pattern on the crop growth in the cultivation racks. Kim et al. [32] pointed out that the increase of the number of cultivation racks would significantly increase the airflow resistance and thus reduce the uniformity of temperature and humidity in the greenhouse. However, there are essential differences between vegetable and mushroom plant in solar greenhouses. In fact, the layout of cultivation racks in planting vegetables cannot meet the efficient production of mushrooms because the fungus will produce a significant amount of heat energy. It is important to ensure a uniform temperature distribution in greenhouses for mushroom cultivation [33]. In the production practice, it is found that the ventilation ability of fungus sticks is not strong due to the too-small spacing between the layers of the cultivation racks, resulting in an excessively high temperature to the disadvantage of mushroom growth. Moreover, in the planning and design of large-scale greenhouse parks, the building distance between adjacent greenhouses also has a significant impact on the ventilation. However, little attention has been given to the clustered solar greenhouse.

In the present research, a mushroom solar greenhouse in China is used as the study object, and the ventilation performance is studied by CFD numerical simulation. Through analyzing the influence of the important parameters of orientation arrangement, horizontal spacing, vertical spacing of the cultivation racks, and the building distance between adjacent greenhouses, the optimal arrangement scheme of cultivation racks and the layout strategy of a clustered greenhouse park can be obtained. The results can provide important theoretical guidance for improving the production quality and land utilization rate of mushroom facilities and provide key technical support for the system design of mushroom solar greenhouses.

2. Theories and Methods

2.1. The Experimental Greenhouse

As shown in Figure 1, the greenhouse used in the experiment is the third generation of energy-saving mushroom solar greenhouse located in the Shenyang area (latitude: 41°49′ N, longitude: 123°34′ E, altitude: 42 m). The orientation is 7° from south to west. The east-west length is 60 m. The span is 10 m. The ridge height is 6 m. The north wall height is 4 m. The horizontal projection length of the rear slope is 1.9 m. The thicknesses of the two side walls and the back wall are both 0.5 m. The outside of each wall is installed with a 0.1 m EPS insulation layer., and the front roof of the greenhouse is 0.15 mm PO film.

Figure 1. Schematic diagram of the mushroom solar greenhouse.

Under continuous sunny weather conditions, the environmental monitoring station based on the internet of things platform is used to measure the external environment, including temperature, humidity, solar radiation, wind speed, and wind direction. These measured parameters act as the initial and boundary conditions of the CFD model. The temperature and humidity measurement instrument in the greenhouse is the RC-4HC hygrothermograph (temperature precision: ±0.5 °C, temperature range: −40 °C~85 °C, humidity precision: ±3% RH, humidity range: 0%~100% RH). The device uses the Em50/R/G data collector as the core system. Sensors measuring different environmental parameters are integrated into the same collector for data collection, and the data are recorded every 10 min. The distribution of measuring points in the greenhouse is as follows: The wind speed measuring points from the north to the south are 0.5 m, 3.5 m, and 6.5 m away from the front foot, and there are four measuring points in total. The data collection interval is set at 10 min. The soil measurement points are arranged at a greenhouse soil depth of 0.1 m plane and are 2 m, 5 m, and 8 m away from the front foot, respectively. There are nine air temperature and humidity measuring points, distributed on four different planes, which are 2 m, 5 m, and 7 m away from the front roof, and the measuring planes are 1 m, 3 m, 5 m, and 6 m away from the ground. The time interval of data collection is 10 min. The temperature and humidity sensors, positioned 10 cm below the soil, are numbered A1–A3 order. Similarly, the height of 1 m is numbered A4–A6. The height of 3 m is numbered A7–A9. The height of 5 m is numbered A10 and A11, and the height of 6 m is numbered A12. The specific arrangement of measuring points is shown in Figure 2.

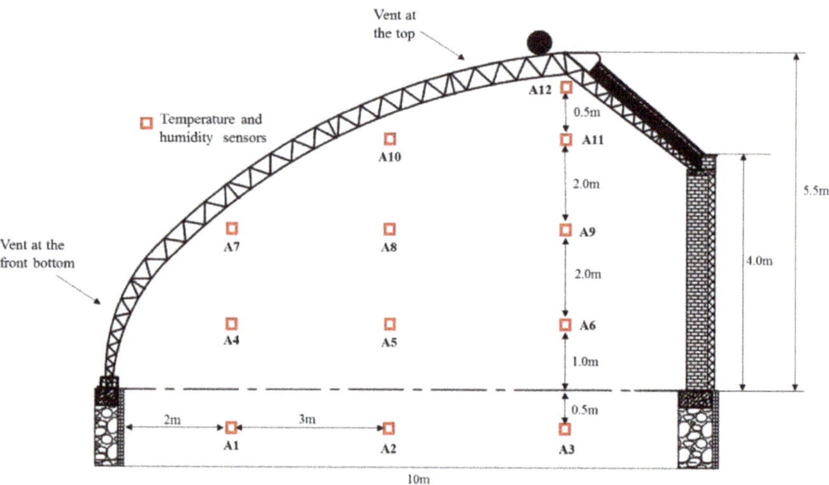

Figure 2. Distribution of measuring points in the experimental testing.

2.2. Geometry Model and Meshing

As shown in Figure 3a, the 3-D geometry model is built according to the actual structure parameters of a mushroom solar greenhouse. As displayed in Figure 3b, the computational domain is divided by an unstructured hexahedral grid, and the size threshold is set as 25 mm. Moreover, grid refinement is carried out for the vents of the solar greenhouse, where the threshold is set as 3 mm. The mesh quality is controlled based on the mesh distortion standard. The maximum distortion range is 0.42~0.66. The weight is 0.24. The mean value is 0.52, and the standard deviation is 0.07. The mesh number of each model ranges from 4.68×10^5 to 4.77×10^5. According to the verification of mesh independence and maximum deformation, the numerical simulation requirements can be achieved.

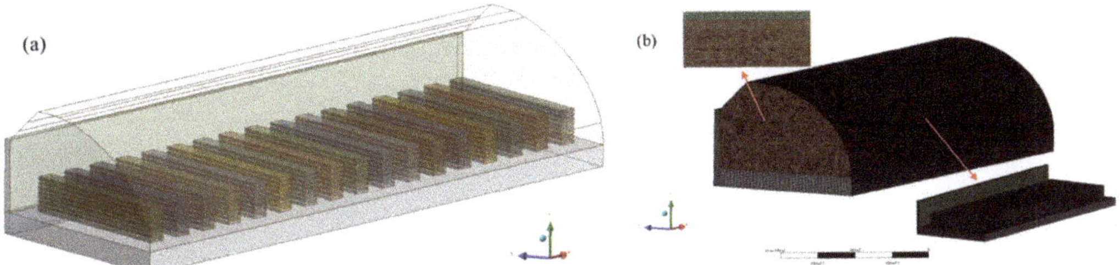

Figure 3. Geometry and grid model of the mushroom solar greenhouse: (**a**) geometric configuration; (**b**) grid division.

2.3. Governing Equations

The fluid dynamics mainly follow the mass conservation equation, energy conservation equation, and momentum conservation equation. The finite volume method is used for the numerical simulation of the mushroom solar greenhouse. The computational domain is divided into a finite number of single, small discretization units to solve control equations. By this means, the physical parameters of each node can be obtained to effectively analyze the fluid flow. The mass conservation equation represents the volume and mass changes caused by the density changes of substances in unit space in unit time, and its expression is as follows:

$$\frac{\partial \rho}{\partial t} + \nabla\left(\rho \vec{v}\right) = 0 \tag{1}$$

where ρ is the fluid density (kg·m^{-3}), t is the time (s), and $\vec{v}$ is the velocity vector.

The momentum conservation equation is the accumulated amount of all external forces acting on the computing unit in time, namely, the momentum changes of the computing unit. The expression is as follows:

$$\frac{\partial}{\partial t}\left(\rho \vec{v}\right) + \nabla\left(\rho \vec{v} \vec{v}\right) = -\nabla P + \nabla \cdot \left(\overline{\overline{\tau}}\right) + \rho \vec{g} \tag{2}$$

where P is the pressure, $\overline{\overline{\tau}}$ is the stress tensor, and $\vec{g}$ is the gravity acceleration.

The energy conservation equation is the spatial accumulation of the resultant force on the computing unit, namely, the energy change of the computing unit. The expression is as follows:

$$\frac{\partial}{\partial t}(\rho E) + \nabla \cdot \left(\vec{v}(\rho E + P)\right) = \nabla \cdot \left(k_{eff} \nabla T - \sum_j h_j \vec{J}_j + \left(\overline{\overline{\tau}}_{eff} \cdot \vec{v}\right)\right) \tag{3}$$

where E is the flow energy, k_{eff} is the effective conductivity, T represents the temperature, h_j is the sensible enthalpy, $\overline{\overline{\tau}}_{eff}$ is the effective viscosity shear, and $\vec{J}_j$ is the diffusion flux of species.

The discrete coordinate (DO) model can be directly applied to transparent, translucent, and opaque optical media such as air, film, glass, and soil surface. It has been used in many climate models of greenhouses [34]. In this numerical simulation, the DO model is employed to calculate the radiation heat flux in several directions.

In order to simulate the internal environment in solar greenhouses, different turbulence models have been compared [35]. The results indicate that the standard k-ε turbulence model has the highest prediction accuracy. Therefore, the standard k-ε turbulence model is selected in this research, and the expressions of the k equation and ε equation are as follows:

$$\frac{\partial}{\partial t}(\rho k)+\frac{\partial}{\partial x_i}(\rho k u_i)=\frac{\partial}{\partial x_j}\left[\left(\mu+\frac{\mu_t}{\sigma_k}\right)\frac{\partial k}{\partial x_j}\right]+G_b+G_k-Y_M-\rho\varepsilon \tag{4}$$

$$\frac{\partial}{\partial t}(\rho \varepsilon)+\frac{\partial}{\partial x_i}(\rho \varepsilon u_i)=\frac{\partial}{\partial x_j}\left[\left(\mu+\frac{\mu_t}{\sigma_\varepsilon}\right)\frac{\partial \varepsilon}{\partial x_j}\right]+C_{1\varepsilon}\frac{\varepsilon}{k}G_k-C_{2\varepsilon}\rho\frac{\varepsilon^2}{k} \tag{5}$$

where k is the turbulence kinetic energy, ε is the dissipation rate of turbulence kinetic energy, u_i is the velocity vector in Cartesian coordinates, μ_t is the eddy viscosity, and σ_k and σ_ε are the turbulent Prandtl numbers; their values are $\sigma_k = 1.0$ and $\sigma_\varepsilon = 1.3$, respectively. $C_{1\varepsilon}$ and $C_{2\varepsilon}$ are the model courants, and their values are $C_{1\varepsilon} = 1.44$ and $C_{2\varepsilon} = 1.92$, respectively. G_k is the generation of k due to mean velocity gradients, G_b is the generation of k due to buoyancy, and Y_M is the contribution of fluctuating dilation in compressible turbulence to overall dissipation rate.

2.4. Numerical Details

In summer, the south wind is dominant in Shenyang, China. Therefore, the south wind direction is set as 0° angle, and the wind speed is set as 5 m/s according to the actual situation. The airflow in the greenhouse is very complicated and not only has a large range but also produces many eddy currents and a large Reynolds coefficient. In the established model, the east, west, and north sides of the greenhouse are set as opaque wall boundaries, while the south roof is defined as a translucent boundary. The physical parameters of soil, envelope, and air in the numerical model are shown in Table 1.

Table 1. Material property parameters in the mushroom solar greenhouse.

Materials	Density ($kg \cdot m^{-3}$)	Heat Conductivity Coefficient ($W \cdot m^{-1} \cdot K^{-1}$)	Specific Heat ($J \cdot kg^{-1} \cdot K^{-1}$)
Air	1.22	0.0242	1006.43
PE plastic film	923	0.38	2300
EPS	9	0.032	1246
Soil	1900	2.0	2200

The 3-D transient analytical solution is calculated in ANSYS FLUENT. The discrete format is the second-order upwind format, and the semi-implicit method for the pressure linked equation (SIMPLE) algorithm is used as a means of pressure and velocity correction. The absolute convergence criterion is 10^{-6} for the energy equation, 10^{-3} for the mass and momentum equation, and 10^{-4} for the other equations. The report function is used to calculate the ventilation cooling efficiency and wind speed.

2.5. Experimental Verification

The CFD simulation results are compared with the experimental measurements to verify the prediction accuracy of the established model. As shown in Figure 4, the comparison of the measuring points at different heights is basically consistent, which indicates that the numerical simulation shows good agreement with the experiment. The maximum relative error between the average simulated and the measured values is at 5.5 m from the ground. Because this height is relatively high in the greenhouse, it is easy for there to be gaps in the film in the ventilation position near the top. Therefore, the temperature at the measurement point A12 is affected by the airflow outside the greenhouse, and the relative error is 5.42%. In the soil at the height of −0.1 m, the difference between the average simulated and measured values is 0.94 °C, with a relative error of 5.08%. This is because the front foot and top vents of the greenhouse are easily affected by external airflow.

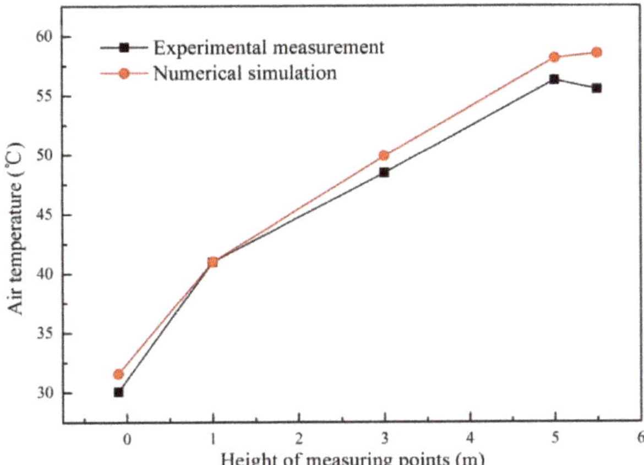

Figure 4. Comparison between the numerical simulation and experimental measurement.

3. Results

3.1. Effect of Orientation Arrangement of Cultivation Racks

A model is established for two arrangements (i.e., north-south vertical arrangement and east-west horizontal arrangement) of cultivation racks frequently used by farmers in production practice, and the influence of different orientation modes on the ventilation performance of fungus sticks is analyzed. In the premise of ensuring the same fungus sticks, the specific parameters are listed in Table 2.

Table 2. Parameters of cultivation racks in the orientation arrangement investigation.

	North-South Vertical Arrangement	East-West Horizontal Arrangement
Length (m)	8	58
Number (row)	29	4
Height (cm)	25	25

Through numerical simulation methods, Figure 5 demonstrates the distribution of wind speed at different heights (i.e., 0.5 m, 1 m, 1.5 m) of the mushroom solar greenhouse under the two orientation modes of cultivation racks. The airflow field inside the greenhouse reached a stable state at 50 s after the vent is opened. As can be seen from the velocity distribution, the wind speed near the ground of the greenhouse reaches the maximum, while that on the cross section shows a decreasing trend as the height increases. In the east-west arrangement mode, the maximum wind speeds on the sections of 0.5 m, 1 m, and 1.5 m are 8.2 m/s, 7.0 m/s, and 4.7 m/s, respectively. Nevertheless, in the north-south arrangement mode, they are 6.9 m/s, 4.7 m/s, and 2.3 m/s, respectively. There are two important findings. On the one hand, the wind speed at each height in the vertical arrangement mode is larger than that in the horizontal arrangement mode. On the other hand, the wind speed distribution of the vertical arrangement is significantly more uniform than that of the horizontal arrangement. The reason for this can be explained by the ventilation effect. The comprehensive ventilation of front bottom and top is adopted in the mushroom solar greenhouse. Therefore, the vertical arrangement of cultivation racks is conducive to providing smooth passage for airflow propagation. As a result, the ventilation efficiency of fungus sticks is higher.

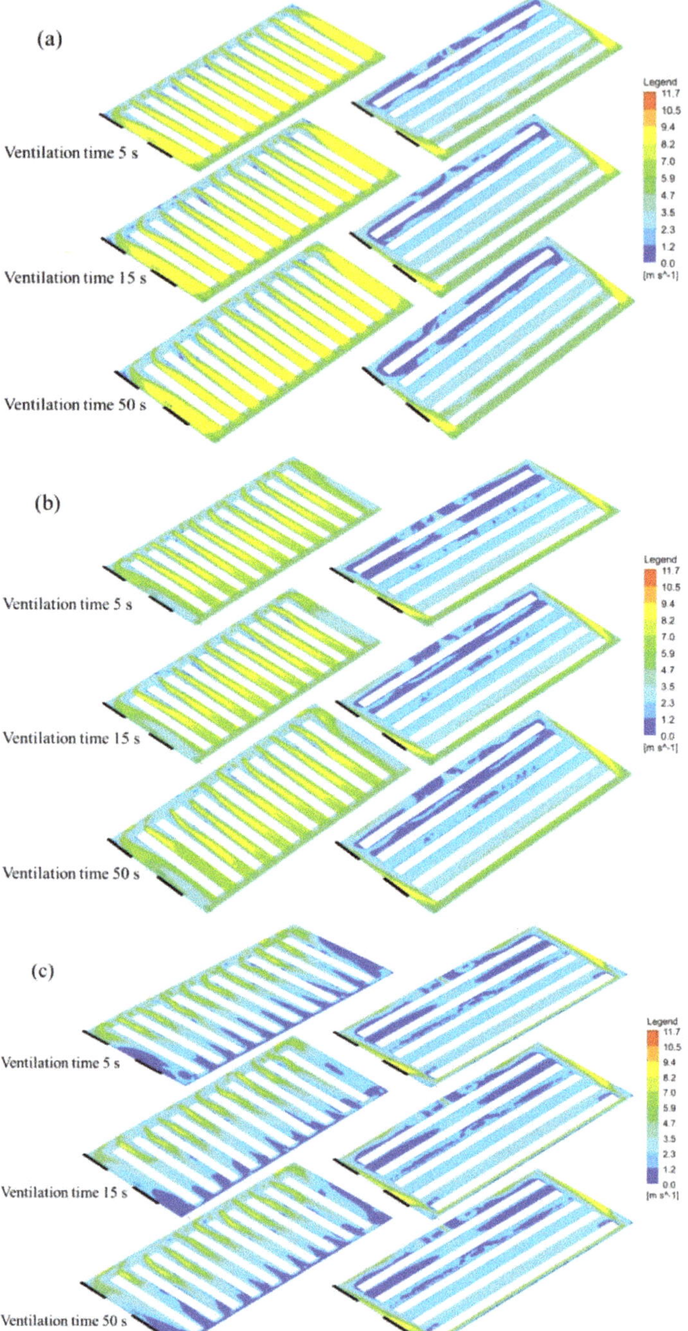

Figure 5. Velocity distributions of typical horizontal cross sections under different arrangements of cultivation racks: (**a**) 0.5 m height; (**b**) 1.0 m height; (**c**) 1.5m height.

Figure 6 shows the comparison of average wind velocity under the two arrangement modes of cultivation racks. On each cross section, the wind speeds of the east-west arrangement at three different heights are significantly higher than that of the north-south arrangement. The average velocity of the east-west arrangement at 0.5 m height is 2.4 m/s higher than that of the north-south arrangement. Similarly, the average velocities of the east-west arrangement at 1.0 m and 1.5 m are 1.4 m/s and 0.8 m/s higher than that of the north-south vertical arrangement, respectively. The results indicate that the north-south arrangement of cultivation racks significantly improves the ventilation performance of the mushroom fungus sticks.

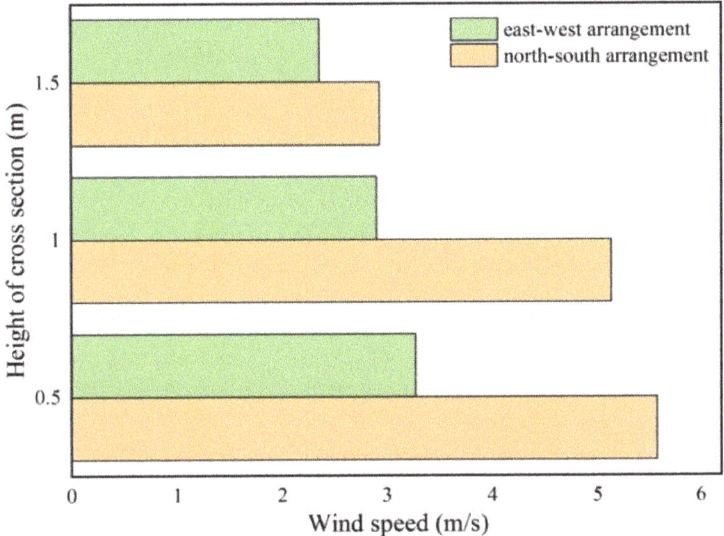

Figure 6. Velocity magnitudes of typical horizontal cross sections under different arrangements of cultivation racks.

The speed vectors under the two modes were analyzed, and the numerical simulation results are illustrated in Figure 7. Due to the blocking effect of the east-west cultivation racks in the front row, the fresh air from outside cannot be smoothly blown to the rear cultivation racks, but can only be blown through the small gap between the layers. As a result, the ventilation effect on the cultivation racks in the last row cannot be achieved. Nevertheless, in the north-south arrangement of cultivation racks, the outside wind can smoothly flow into the greenhouse without the blocking effect. The cooling rate of each layer of the cultivation rack is reasonable. The wind can form a uniform flow in every gap between two rows of cultivation racks, which guarantees the ventilation efficiency in the mushroom solar greenhouse.

Statistical analysis was conducted on the average velocity of different heights, and the processing results are shown in Figure 8. It can be seen that the wind speed of the north-south arrangement in most areas of the greenhouse is much larger than that of the east-west arrangement. However, the wind speed of the east-west arrangement is slightly larger at the 1/3 section of the greenhouse. This is because the horizontally arranged cultivation racks in the front rows will block the wind flow into the greenhouse. The airflow has to move from the gap near the film to the top location of the greenhouse. As a result, fresh wind that has not been fully exchanged will escape from the top vent. What is worse, is that the last two rows have little airflow because the fresh wind cannot smoothly reach those cultivation racks. In conclusion, the appropriate orientation arrangement of cultivation racks in the mushroom solar greenhouse is the north-south arrangement. The average wind velocity in this layout is larger in comparison with the east-west arrangement, and the airflow is more uniform. It not only helps improve the ventilation efficiency in each layer of the cultivation racks, but also contributes to the breath of mushroom fungus sticks.

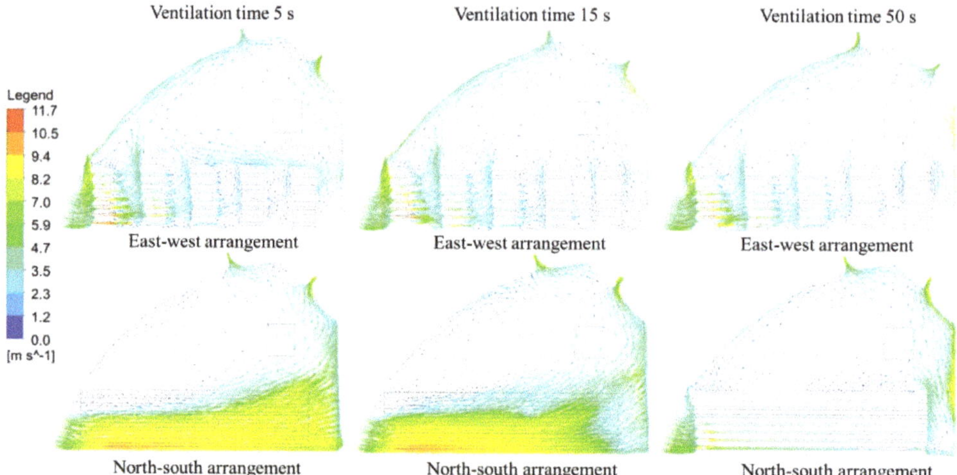

Figure 7. Velocity vectors of typical vertical cross sections under different arrangements of cultivation racks.

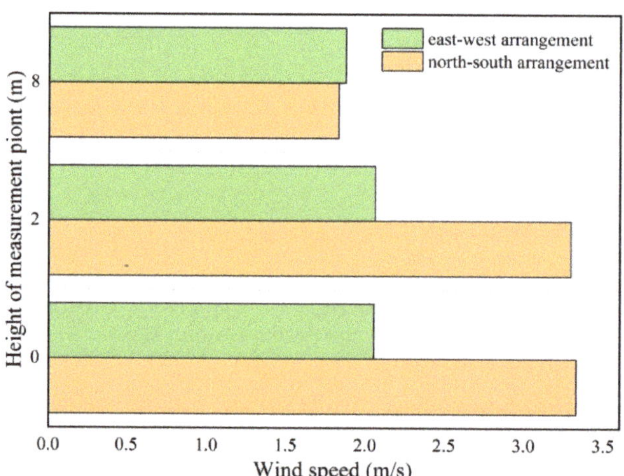

Figure 8. Velocity magnitudes of typical vertical cross sections under different arrangements of cultivation racks.

3.2. *Effect of Horizontal Spacing of Cultivation Racks*

As the configuration of cultivation racks in the mushroom solar greenhouse does not have a definitive design, the present research has carried out numerical simulation to determine the horizontal spacing of the cultivation racks. In order to find out the best horizontal spacing, seven different cases were investigated, and the detailed parameters are exhibited in Table 3.

Table 3. Parameters of cultivation racks in the horizontal spacing investigation.

Order	A	B	C	D	E	F	G
Horizontal spacing (m)	0.6	0.7	0.8	0.9	1.0	1.1	1.2
Racks number	20	19	18	17	16	15	15

The ventilation characteristics of different horizontal spacings of cultivation racks were compared, and the numerical results are displayed in Figure 9. The numerical results indicate that the overall velocity distribution is not significantly affected by the horizontal spacing. Nevertheless, the magnitude of the wind speed is distinguishing. As can be seen from the horizontal spacing of 0.6 m and 0.7 m, the maximum wind speed between adjacent cultivation racks reaches 3.21 m/s due to the small spacing. Nevertheless, most of the airflow directly moves in the corridor between the cultivation racks, resulting in low wind speed and uneven velocity distribution inside the cultivation racks. When the spacing is enlarged from 0.8 m to 1.2 m, the wind can run through the internal fungus stick of the cultivation racks, so as to achieve the desired ventilation effect. Through the numerical simulation, it can be concluded that the larger the spacing, the more advantageous it is for the ventilation inside the greenhouse. However, the horizontal spacing between the existing cultivation racks is too low. Some of the cultivation racks near the back wall do not have the fresh wind evenly blowing to them. As the horizontal spacing of the cultivation racks is enlarged to 1.2 m, the predicament is relieved. Nevertheless, too-large horizontal spacing is detrimental to the scale of production. In this way, the number of fungus sticks will be reduced, which will greatly reduce the total output of the mushroom solar greenhouse.

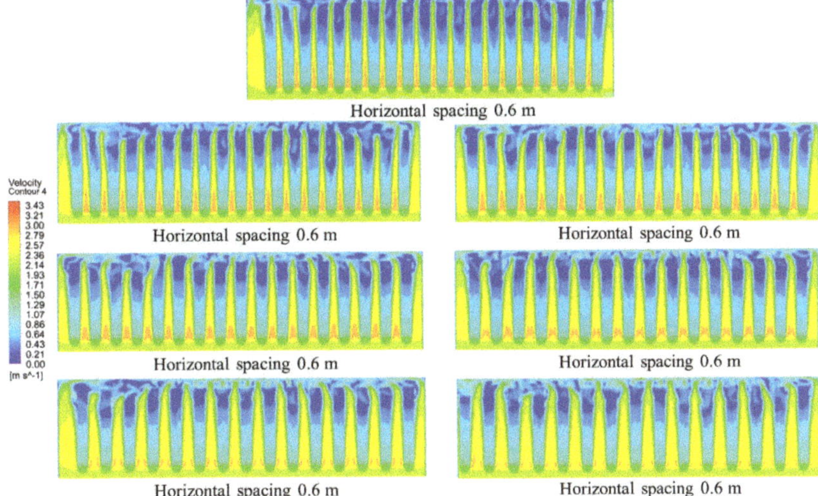

Figure 9. Velocity distributions of typical cross sections under the different horizontal spacings of cultivation racks.

Statistical analysis was conducted on the average wind speed under different horizontal spacings of the cultivation racks, and the results are demonstrated in Figure 10. The average airflow velocities of the five cases with horizontal spacing of 0.8 m, 0.9 m, 1.0 m, 1.1 m, and 1.2 m, all of which are 0.65 m/s, are basically the same. However, the wind speeds of 0.6 m and 0.7 m are about 0.46 m/s when the external wind just enters the greenhouse, and the wind speeds only reach about 0.6 m/s when the ventilation period is 15 s. This indicates that too-small horizontal spacing is to the disadvantage of ventilation of the cultivation racks. The difference of wind speed between 0.8 m and 1.2 m is only 0.05 m/s, suggesting that the ventilation performance of the cultivation racks can be satisfied when the horizontal spacing is 0.8 m. This is consistent with the above results.

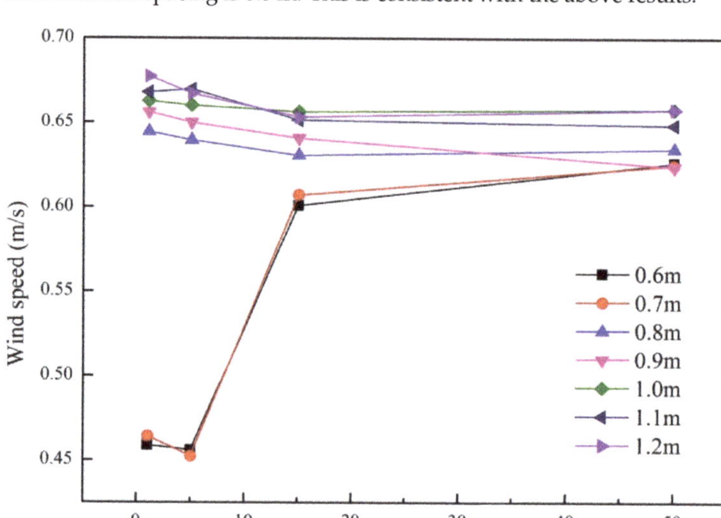

Figure 10. Velocity magnitudes of typical cross sections under different horizontal spacings of cultivation racks.

3.3. Effect of Vertical Spacing of Cultivation Racks

There is no systematic study on the ventilation effect of the spacing between cultivation rack layers (i.e., vertical spacing) on the ventilation performance. Based on the above research results, the layout of cultivation racks is the north-south arrangement. The horizontal spacing is specified as 0.8 m, and the vertical spacing of the cultivation racks is investigated with an interval of 0.1 m and 0.01 m for interpolation modeling. The specific parameters of the six different established cases are shown in Table 4.

Table 4. Parameters of cultivation racks in the vertical spacing investigation.

Order	A	B	C	D	E	F
Vertical spacing (cm)	26	27	28	29	30	31
Height of cultivation rack (m)	2.18	2.26	2.34	2.42	2.50	2.58

Numerical simulation was conducted on the overall ventilation rate of the greenhouse under different vertical spacings of the cultivation racks, and the results are illustrated in Figure 11. It can be seen from the velocity distribution that the wind speed is more uniform with an increase in the vertical spacing between adjacent cultivation racks. Moreover, the greater the vertical spacing between the layers is, the larger the average wind speed is. When the vertical spacing is 26 cm, the wind speed inside the cultivation racks near the back wall is very low. Only the front part has airflow, which is too high and reaches 2.48 m/s. With the increase of vertical spacing, the cultivation racks near the back wall gradually acquire wind flow. When the vertical spacing increases to 29 cm, the whole cultivation rack can gain access to wind flow, and the distribution of wind speed is much more uniform. As the vertical spacing of the cultivation racks increase to 30 cm and 31 cm, however, the ventilation performance is disappointing. Even though the wind can run through the whole cultivation rack, the distribution of wind speed is disorganized.

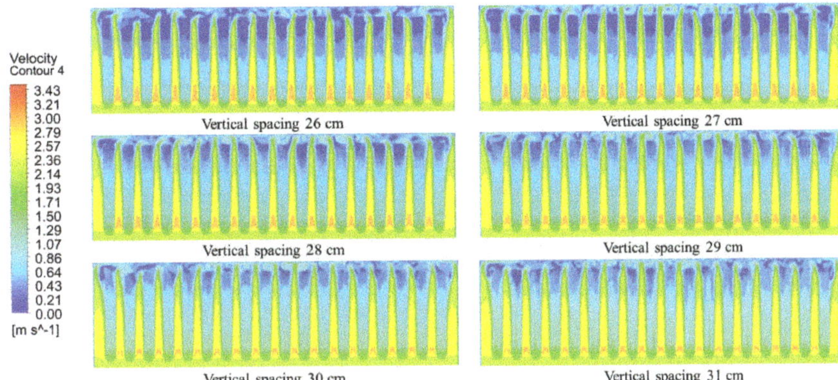

Figure 11. Velocity distributions of typical cross sections under different vertical spacings of cultivation racks.

Figure 12 exhibits the statistical value of the average wind speed in each cross section. The results indicate that the average velocity first increases and then decreases according to the increase in the vertical spacing from 26 cm to 31 cm. and the maximum wind speed reaches 1.26 m/s at 29 cm. If the vertical spacing is increased further, the ventilation effect becomes weak. Therefore, the best vertical spacing of the cultivation rack layer is 29 cm.

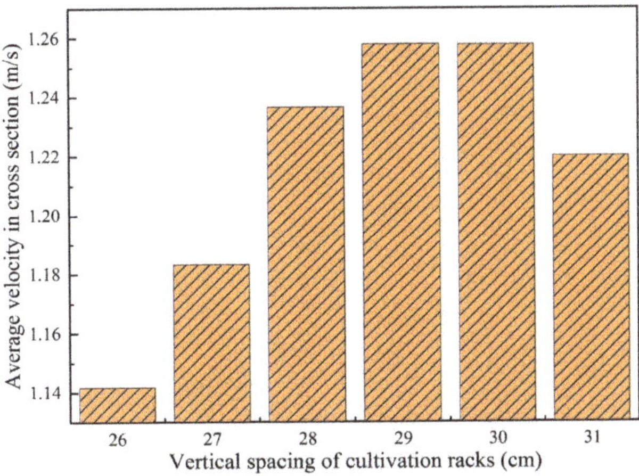

Figure 12. Velocity magnitudes of typical cross sections under different vertical spacings of cultivation racks.

3.4. Ventilation Design of Clustered Greenhouse Park

The simulation model of twelve greenhouses in a double row has been established in the cluster greenhouse park, and the ventilation effect is displayed in Figure 13. It can be seen that the airflow outside the first three rows of greenhouses is irregular, and the wind speed at the air inlet of each greenhouse is inconsistent. Nevertheless, the airflow movement of the front inlet of each greenhouse is relatively consistent from the fourth row to the sixth row, which is quite different from the first three greenhouses. By contrast, the wind speed of the middle cross section in the six rows is uniform without significant difference. It can be clearly seen from the top cross section that the airflow in the first two rows is irregular, and the wind speed at the outlet of each greenhouse is inconsistent. However, the airflow movement of each greenhouse is relatively consistent from the third row to the sixth row.

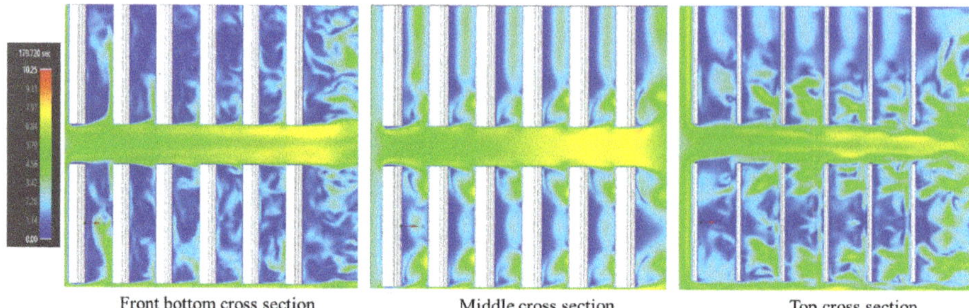

Figure 13. Flow fields of three typical cross sections under different greenhouse arrangement.

Furthermore, in order to reduce the calculation cost and to improve the accuracy, an enriched model of three greenhouses was established. The ventilation and cooling efficiency of different building distances before and after the greenhouse were studied. Because there is no shading effect in the front greenhouse, this numerical simulation mainly analyzes the second and third row of greenhouses. Figure 14 shows the comparison of cooling efficiency under different building spacings between the two greenhouses. Figure 14a demonstrates the internal temperature variation inside the second row under the building spacing from 5 m to 14 m. As can be seen, the internal cooling rate of the greenhouse decreases obviously with the increase in building spacing. It shows a decreasing trend of 0.3 m/s for every 1 m increase. However, the quantitative change occurs between 10 m and 11 m. The cooling rate of 11 m spacing is significantly faster than that of 10 m from the beginning of 10 s, and the gap increases to 1 °C in the ventilation period of 40 s. However, the difference fades away after 100 s. Moreover, the cooling rate does not change significantly when the building spacing increases to 12 m and 14 m. The cooling efficiency of the third row of clustered greenhouses was also analyzed, and the results are shown in Figure 14b. It can be seen that the cooling trend of the third row is consistent with that of the second row. The temperature decreases by 0.1 °C every 1 m increase in the building spacing. The third row also undergoes quantitative changes at 10 m and 11 m. In the ventilation period of 20 s, the cooling rate is significantly faster when the building spacing is 11 m than when it is 10 m, and the maximum temperature difference reaches to 0.8 °C at 50 s. Nevertheless, the temperature at 10 m and 11 m is basically the same when the ventilation time reaches 100 s, and there is no change in the cooling rate when the building spacing increases to 12 m and 14 m.

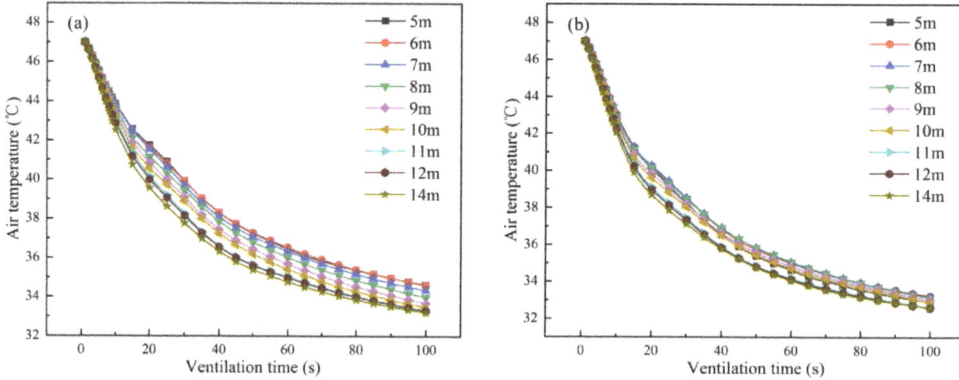

Figure 14. Cooling rate of the clustered solar greenhouse under different building spacings: (**a**) the second row; (**b**) the third row.

The ventilation performance was studied, and the distribution of the average wind speed in the clustered greenhouses is exhibited in Figure 15. The wind speed inside the second row is analyzed in Figure 15a. According to intuition, the greater the building spacing between greenhouses is, the greater the wind speed. Nevertheless, the velocity magnitude first increases to the maximum, and then slowly reduces to smooth. The average wind speed rises gradually with the increase in building spacing from 5 m to 7 m. However, the change of airflow is not significant from 8 m to 14 m. The peak value of wind speed reaches 0.71 m/s under the building spacing of 10 m. After the ventilation flow is stable in the greenhouse, the wind speed at 11 m, 12 m, and 14 m are 0.59 m/s, 0.62 m/s, and 0.68 m/s, respectively. The fungus sticks actually need high wind speed to achieve efficient ventilation. In consideration of the growth habit of mushroom, the optimal building spacing of the second row of clustered greenhouses is 10 m. Corresponding analysis was carried out on the third row, and the results are shown in Figure 15b. The wind speed increases to a maximum, and then slowly reduces to smooth. The velocity magnitude gradually increases as the building spacing is enlarged from 5 m to 9 m. However, there is no significant change from 10 m to 14 m. When the building distance between the clustered greenhouses is 14 m, the peak value of airflow velocity inside the greenhouses is 0.70 m/s. Nevertheless, it falls rapidly to 0.58 m/s when the ventilation effect reaches a stable level at 100 s. The average wind speed under the building spacing of 11 m, 12 m, and 14 m are 0.49 m/s, 0.52 m/s, and 0.54 m/s, respectively. With the increase in ventilation time, the wind speed inside the greenhouse begins to rise until it reaches a stable level when the building spacing is 10 m. Therefore, the optimal building spacing of the third row of greenhouses is also 10 m.

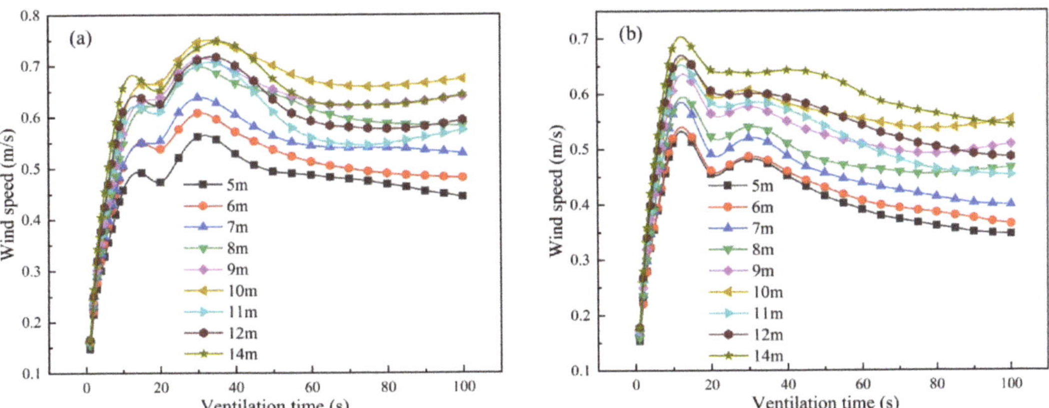

Figure 15. Air velocity variation of the clustered solar greenhouse under different building spacings: (**a**) the second row; (**b**) the third row.

4. Discussion

The growth of mushrooms is greatly affected by environmental factors, such as temperature, humidity, illumination, and CO_2 concentration. Among these environmental factors, the most influential one is the temperature. A too-high or too-low temperature will cause damage to the growth of mushrooms. The optimal growth temperature is 5~33 °C for mycelial growth and 12~28 °C for mushroom emergence. Because the growth of mushrooms requires less solar radiation intensity, the ventilation plays a very important role in the exchange of micro-environment and energy in the production process of solar greenhouses. In some places, it is difficult to grow mushrooms using conventional methods (i.e., cold winter or strong sunshine in summer, etc). However, in solar greenhouses, the mushroom can be grown in some harsh climate conditions. Previous scholars have performed a series of investigations into vegetable greenhouses and have obtained important results with academic value. However, little attention has been paid to the system design of a mushroom solar greenhouse. On the one hand, the mushroom has more demanding requirements on temperature and humidity; therefore, the ventilation results of a vegetable greenhouse cannot be directly applied to a mushroom greenhouse. On the other hand, the fungus sticks are produced in the form of cultivation racks, in which the ventilation is much more complicated than in soil cultivation. Nevertheless, there is a research gap in the configuration scheme of cultivation racks such as orientation arrangement, horizontal spacing, and vertical spacing. In addition, the planning and design of the vegetable greenhouse park is based on the principle that the front greenhouse does not affect the lighting of the back greenhouse. However, it does not fit into the weak light habit of mushroom. There is no systematic study on the ventilation of a large-scale park of mushroom solar greenhouses. In order to improve the production efficiency in the clustered mushroom greenhouse park, it is of significant importance to establish the theoretical basis of ventilation design theory.

In order to explore the system design of clustered mushroom solar greenhouses, the present research has developed a 3-D mathematical model based on CFD theory to study the ventilation performance. According to the analysis of different orientation arrangements of cultivation racks, it can be asserted that the best design is the north-south arrangement. This arrangement not only maximizes the wind speed in the greenhouse, but also enables the fresh air into the greenhouse to flow evenly to each cultivation shelf, which is conducive to the ventilation of fungus sticks. In addition, through the analysis of the horizontal spacing of the cultivation racks, it can be concluded that the optimal horizontal spacing is 0.8 m. If the separation distance of the cultivation racks is blindly increased, the overall airflow velocity in the greenhouse will not improve significantly, but will waste the land utilization. According to the analysis of vertical spacing of cultivation racks, it is suggested that the optimal vertical spacing of cultivation racks is 29 cm. This configuration helps create a smooth passage to achieve the best ventilation effect of the fungus sticks. According to the analysis of the cooling rate and ventilation efficiency of clustered greenhouses, the optimal building spacing is 10 m. This arrangement contributes to the normal growth and development of mushroom fungus sticks, which can not only increase the utilization rate of land, but also ensure the best cooling rate of each greenhouse.

Greenhouse cultivation is a suitable agricultural techniques for creating and controlling the inside microclimate to be adequate for plant growth. For mushroom solar greenhouses in summer, the temperature is too high to cause the normal growth of fungus crops under the strong solar radiation and weak wind. It is of great significance to find the reasonable planning of clustered greenhouse parks and the effective method of rapid cooling inside the greenhouse to ensure the high-quality production. As a matter of fact, the major environmental factors, including temperature, humidity, illuminance, and gas concentration, in the greenhouse are invisible. Nevertheless, the reported research on greenhouse ventilation is still mainly focused on experimental research, which has long cycle, high cost, and is difficult to reproduce according to different climate conditions. The CFD numerical simulation can effectively solve the above problems and can realize the visual analysis of airflow field and temperature distribution. However, there are few reports on the numerical simulation of mushroom solar greenhouse ventilation. The following work will further use CFD to study the effect of vent configuration on the ventilation performance of mushroom solar greenhouses.

5. Conclusions

In this study, the CFD theory has been used to establish a mathematical model for the numerical simulation of clustered mushroom solar greenhouses. The effects of orientation arrangement, horizontal spacing, vertical spacing of the cultivation racks, and the building spacing between clustered greenhouses on the ventilation performance have been investigated. The best orientation arrangement of cultivation racks inside the mushroom solar greenhouse is a north-south layout because of the maximum wind speed. Moreover, the fresh wind into the greenhouse can flow evenly to each layer of cultivation racks, which contributes to the ventilation of mushroom fungus sticks. The optimal horizontal spacing of the cultivation racks is 0.8 m, which is conducive to wind running through the interior of the cultivation racks, so as to achieve a reasonable ventilation effect. The optimal vertical spacing between cultivation rack layers is 29 cm, which can not only avoid the shielding effect, but can also improve the uniformity of wind speed distribution. For the clustered mushroom solar greenhouse park, the best building spacing between adjacent greenhouses is 10 m, which can not only increase the land utilization rate, but also ensure the best cooling effect of each greenhouse.

Author Contributions: Conceptualization, Y.L. and X.L.; methodology, Y.L. and X.L.; software, F.S.; validation, Y.L., W.S. and T.L.; formal analysis, F.S. and W.S.; investigation, F.S.; resources, W.S.; data curation, Y.L.; writing—original draft preparation, Y.L. and W.S.; writing—review and editing, X.L. and T.L.; visualization, Y.L.; supervision, X.L.; project administration, T.L.; funding acquisition, Y.L. and T.L. All authors have read and agreed to the published version of the manuscript.

Funding: This research was funded by the China Postdoctoral Foundation of Yiming Li (No. 275737) and the China Agriculture Research System of MOF and MARA (CARS-23-C01).

Institutional Review Board Statement: Not applicable.

Informed Consent Statement: Not applicable.

Data Availability Statement: Not applicable.

Conflicts of Interest: The authors declare no conflict of interest.

References

1. Tong, G.; Christopher, D.M.; Li, B. Numerical modelling of temperature variations in a Chinese solar greenhouse. *Comput. Electron. Agric.* **2009**, *68*, 129–139. [CrossRef]
2. Majdoubi, H.; Boulard, T.; Fatnassi, H.; Bouirden, L. Airflow and microclimate patterns in a one-hectare canary type greenhouse: An experimental and CFD assisted study. *Agric. For. Meteorol.* **2019**, *149*, 1050–1062. [CrossRef]
3. Li, H.; Li, Y.; Yue, X.; Liu, X.; Tian, S.; Li, T. Evaluation of airflow pattern and thermal behavior of the arched greenhouses with designed roof ventilation scenarios using CFD simulation. *PLoS ONE* **2020**, *15*, e0239851. [CrossRef]
4. Liu, X.; Wu, X.; Xia, T.; Fan, Z.; Shi, W.; Li, Y.; Li, T. New insights of designing thermal insulation and heat storage of Chinese solar greenhouse in high latitudes and cold regions. *Energy* **2022**, *242*, 122953. [CrossRef]
5. Zhang, X.; Wang, H.; Zou, Z.; Wang, S. CFD and weighted entropy based simulation and optimisation of Chinese Solar Greenhouse temperature distribution. *Biosyst. Eng.* **2016**, *142*, 12–26. [CrossRef]
6. Choab, N.; Allouhi, A.; Maakoul, A.E.; Kousksou, T.; Saadeddine, S.; Jamil, A. Review on greenhouse microclimate and application: Design parameters, thermal modeling and simulation, climate controlling technologies. *Sol. Energy* **2019**, *191*, 109–137. [CrossRef]
7. Xu, F.; Shang, C.; Li, H.; Xue, X.; Sun, W.; Chen, H.; Li, Y.; Zhang, Z.; Li, X.; Guo, W. Comparison of thermal and light performance in two typical Chinese solar greenhouses in Beijing. *Int. J. Agric. Biol. Eng.* **2019**, *12*, 24–32. [CrossRef]
8. Wu, X.; Liu, X.; Yue, X.; Xu, H.; Li, T.; Li, Y. Effect of the ridge position ratio on the thermal environment of the Chinese solar greenhouse. *R. Soc. Open Sci.* **2021**, *8*, 201707. [CrossRef]
9. Tong, G.; Christopher, D.M.; Zhao, R.; Wang, J. Effect of location and distribution of insulation layers on the dynamic thermal performance of Chinese solar greenhouse walls. *Appl. Eng. Agric.* **2014**, *30*, 457–469.
10. Cao, K.; Xu, H.; Zhang, R.; Xu, D.; Yan, L.; Sun, Y.; Xia, L.; Zhao, J.; Zou, Z.; Bao, E. Renewable and sustainable strategies for improving the thermal environment of Chinese solar greenhouses. *Energy Build.* **2019**, *202*, 109414. [CrossRef]
11. Ma, Y.; Li, X.; Fu, Z.; Zhang, L. Structural design and thermal performance simulation of shade roof of double-slope greenhouse for mushroom-vegetable cultivation. *Int. J. Agric. Biol. Eng.* **2019**, *12*, 126–133. [CrossRef]
12. Zheng, D.; Wang, J.; Chen, Z.; Baleta, J.; Sundén, B. Performance analysis of a plate heat exchanger using various nanofluids. *Int. J. Heat Mass Transf.* **2020**, *158*, 119993. [CrossRef]
13. Wang, J.; Li, G.; Li, T.; Zeng, M.; Sundén, B. Effect of various surfactants on stability and thermophysical properties of nanofluids. *J. Therm. Anal. Calorim.* **2021**, *143*, 4057–4070. [CrossRef]
14. Zheng, D.; Yang, J.; Wang, J.; Kabelac, S.; Sundén, B. Analyses of thermal performance and pressure drop in a plate heat exchanger filled with ferrofluids under a magnetic field. *Fuel* **2021**, *293*, 120432. [CrossRef]
15. Wang, J.; Li, G.; Zhu, H.; Luo, J.; Sundén, B. Experimental investigation on convective heat transfer of ferrofluids inside a pipe under various magnet orientations. *Int. J. Heat Mass Transf.* **2019**, *132*, 407–419. [CrossRef]
16. Chen, Z.; Zheng, D.; Wang, J.; Chen, L.; Sundén, B. Experimental investigation on heat transfer characteristics of various nanofluids in an indoor electric heater. *Renew. Energy* **2020**, *147*, 1011–1018. [CrossRef]
17. Amani, M.; Foroushani, S.; Sultan, M.; Bahrami, M. Comprehensive review on dehumidification strategies for agricultural greenhouse applications. *Appl. Therm. Eng.* **2020**, *181*, 115979. [CrossRef]

18. Chan, K.L.; Mo, C.; Shin, K.Y.; Im, Y.H.; Yoon, S.W. A Study of the effects of enhanced uniformity control of greenhouse environment variables on crop growth. *Energies* **2019**, *12*, 091749.
19. McCartney, L.; Lefsrud, M.G. Field trials of the natural ventilation augmented cooling (NVAC) greenhouse. *Biosyst. Eng.* **2018**, *174*, 159–172. [CrossRef]
20. Kittas, C.; Bartzanas, T. Greenhouse microclimate and dehumidification effectiveness under different ventilator configurations. *Build. Environ.* **2007**, *42*, 3774–3784. [CrossRef]
21. Limtrakarn, W.; Boonmongkol, P.; Chompupoung, A.; Rungprateepthaworn, K.; Kruenate, J.; Dechaumphai, P. Computational fluid dynamics modeling to improve natural flow rate and sweet pepper productivity in greenhouse. *Adv. Mech. Eng.* **2012**, *4*, 158563. [CrossRef]
22. Bournet, P.E.; Ould Khaoua, S.A.; Boulard, T. Numerical prediction of the effect of vent arrangements on the ventilation and energy transfer in a multi-span glasshouse using a bi-band radiation model. *Biosyst. Eng.* **2007**, *98*, 224–234. [CrossRef]
23. Villagrá, E.A.; Baeza Romero, E.J.; Bojacá, C.R. Transient CFD analysis of the natural ventilation of three types of greenhouses used for agricultural production in a tropical mountain climate. *Biosyst. Eng.* **2019**, *188*, 288–304. [CrossRef]
24. He, K.; Chen, D.; Sun, L.; Huang, Z.; Liu, Z. Effects of vent configuration and span number on greenhouse microclimate under summer conditions in eastern China. *Int. J. Vent.* **2015**, *13*, 381–396. [CrossRef]
25. Bartzanas, T.; Boulard, T.; Kittas, C. Effect of vent arrangement on windward ventilation of a tunnel greenhouse. *Biosyst. Eng.* **2004**, *88*, 479–490. [CrossRef]
26. Pakari, A.; Ghani, S. Airflow assessment in a naturally ventilated greenhouse equipped with wind towers: Numerical simulation and wind tunnel experiments. *Energy Build.* **2019**, *199*, 1–11. [CrossRef]
27. Zhang, G.; Fu, Z.; Yang, M.; Liu, X.; Dong, Y.; Li, X. Nonlinear simulation for coupling modeling of air humidity and vent opening in Chinese solar greenhouse based on CFD. *Comput. Electron. Agric.* **2019**, *162*, 337–347. [CrossRef]
28. Al-Helal, I.M.; Waheeb, S.A.; Ibrahim, A.A. Modified thermal model to predict the natural ventilation of greenhouses. *Energy Build.* **2015**, *99*, 1–8. [CrossRef]
29. Han, J.H.; Kwon, H.J.; Yoon, J.Y.; Kim, K.; Nam, S.W.; Son, J.E. Analysis of the thermal environment in a mushroom house using sensible heat balance and 3-D computational fluid dynamics. *Biosyst. Eng.* **2009**, *104*, 417–424. [CrossRef]
30. Zhang, Y.; Kacira, M.; An, L. A CFD study on improving air flow uniformity in indoor plant factory system. *Biosyst. Eng.* **2016**, *147*, 193–205. [CrossRef]
31. Baek, M.S.; Kwon, S.Y.; Lim, J.H. Improvement of the crop growth rate in plant factory by promoting air flow inside the cultivation. *Int. J. Smart Home* **2016**, *10*, 63–74. [CrossRef]
32. Kim, K.; Yoon, J.Y.; Kwon, H.J.; Han, J.H.; Son, J.E.; Nam, S.W.; Giacomelli, G.A.; Lee, I.B. 3-D CFD analysis of relative humidity distribution in greenhouse with a fog cooling system and refrigerative dehumidifiers. *Biosyst. Eng.* **2008**, *100*, 245–255. [CrossRef]
33. Ryu, K.J.; Son, J.H.; Han, C.W.; Nah, K.D. A study on the design of air conditioning system in the mushroom cultivation greenhouse. *J. Korea Acad. -Ind. Coop. Soc.* **2017**, *18*, 743–750.
34. Xia, T.; Li, Y.; Wu, X.; Fan, Z.; Shi, W.; Liu, X.; Li, T. Performance of a new active solar heat storage–release system for Chinese assembled solar greenhouses used in high latitudes and cold regions. *Energy Reports.* **2022**, *8*, 784–797. [CrossRef]
35. Nebbali, R.; Roy, J.C.; Boulard, T. Dynamic simulation of the distributed radiative and convective climate within a cropped greenhouse. *Renew. Energy* **2012**, *43*, 111–129. [CrossRef]

MDPI
St. Alban-Anlage 66
4052 Basel
Switzerland
Tel. +41 61 683 77 34
Fax +41 61 302 89 18
www.mdpi.com

Energies Editorial Office
E-mail: energies@mdpi.com
www.mdpi.com/journal/energies

www.ingramcontent.com/pod-product-compliance
Lightning Source LLC
LaVergne TN
LVHW070633170726
843515LV00004B/240

9783036580043